GLOBAL WARMING

Acknowledgments

I would like to thank Johanna Maslin and the Department of Geography, UCL.

This edition published in 2007 by MBI Publishing Company LLC and Voyageur Press, an imprint of MBI Publishing Company, Galtier Plaza, Suite 200, 380 Jackson Street, St. Paul, MN 55101-3885 USA

Voyageur Press titles are also available at discounts in bulk quantity for industrial or sales-promotional use. For details write to Special Sales Manager at MBI Publishing Company, Galtier Plaza, Suite 200, 380 Jackson Street, St. Paul, MN 55101-3885 USA.

To find out more about our books, join us online at www.voyageurpress.com.

ISBN-13: 978-0-7603-2965-8
ISBN-10: 0-7603-2965-6

Printed in China

Graphics on pp 8, 11, 14, 20, 27, 38, 41, 55, 59 by Philippe Rekacewicz, Unep-Grid Arendal, Norway, 2000
Graph on p42 © IPCC, 2001. Reproduced with permission.

Photographs © 2002 by

Front cover: David Nunuk/Science Photo Library
Back cover: Geospace/Science Photo Library
Page 1: Jeremy Walker/Science Photo Library
Page 4: Bernhard Edmaier/Science Photo Library
Page 6: Earth Satellite Corp./Science Photo Library
Page 10: Tim Lester/Science Photo Library
Page 12: BSIP, MIG/BAEZA/Science Photo Library
Page 15: Simon Fraser/Science Photo Library
Page 16: G. Brad Lewis/Science Photo Library
Page 19: Cyril Ruoso/Bios-Still Pictures
Page 22: Ken Biggs/Science Photo Library
Page 23: Bernhard Edmaier/Science Photo Library
Page 24: Jacques Jangoux/Science Photo Library
Page 26: Simon Fraser/Science Photo Library
Page 28: Bernhard Edmaier/Science Photo Library
Page 29: Frans Lemmens/Still Pictures
Page 31 left: Bruno Pambour/Bios-Still Pictures
Page 31 right: Nigel Dickinson/Still Pictures
Page 32: B & C Alexander/Science Photo Library
Page 35: NASA/Science Photo Library
Page 36 top: G. Griffiths-Christian Aid/Still Pictures
Page 36 bottom: Adrian Arbib/Still Pictures
Page 40: Klaus Andrews/Still Pictures
Page 43: Klaus Andrews/Still Pictures
Page 44: Shehzad Noorani/Still Pictures
Page 46 top: Nigel Dickinson/Still Pictures
Page 46 bottom: Jean Roche/Still Pictures
Page 49: Compost/Visage/Still Pictures
Page 51 left: Pascal Kobeh/Still Pictures
Page 51 right: Matthew Oldfield/Scubazoo/Science Photo Library
Page 52 left: Carlos Guarita/Still Pictures
Page 52 right: Mark Edwards/Still Pictures
Page 56: Simon Fraser/Science Photo Library
Page 60: Simon Fraser/Science Photo Library
Page 62: Luiz C. Marigo/Still Pictures
Page 63: NRSC Ltd/Science Photo Library
Page 64: Martin Bond/Science Photo Library
Page 67: Martin Bond/Science Photo Library
Page 68: Chris Knapton/Science Photo Library
Page 69: Colin Baxter. The Blue Lagoon, Reykjanes
From *Iceland* portfolio, page 14
Page 71: Science Photo Library

GLOBAL WARMING

Mark Maslin

WorldLife
LIBRARY

Voyageur Press

Contents

Introduction

Planet Earth is warming faster than at any other time in the past 1000 years and there is little doubt that human activity is to blame. Global warming is due to the massive increase in greenhouse gases, such as carbon dioxide, which we are putting into the atmosphere. Burning of fossil fuel has already raised levels of atmospheric carbon dioxide to their highest for the last 20 million years.

The most recent report by the Intergovernmental Panel on Climate Change (IPCC), shows there is clear evidence for a 1.1°F (0.6°C) rise in global temperatures and a 7 ¾ in (20 cm) rise in sea level during the twentieth century. There is evidence for a 40 per cent reduction in the thickness of sea ice over the Arctic Ocean. Mountain glaciers are melting at the fastest rate ever recorded. There has been a 40 per cent increase in storm activity in the North Atlantic region over the last 50 years and global floods and droughts have become more frequent. In England, the winter of 2000/2001 was the wettest on record, while the heat wave in 2003 killed at least 35,000 people in Europe. The IPCC report predicts that global temperatures will rise by up to 10.4°F (5.8°C) by 2100.

This book explains what global warming is and the evidence that it is really happening. It also examines the devastating effects it will have on human society, the natural environment, and the world economy, including drastic changes in health, agriculture, water resources, coastal regions, storminess, forests and wildlife. For each of these areas scientists and social scientists have made estimates of the potential consequences, and the most important ones will be discussed in this book. This book also discusses the potential surprises the climate system may have in store for us, due to the burning of the Amazon rainforest, the switching off of deep ocean circulation and the possible release of gas hydrates which are currently trapped beneath the ocean. Be under no illusion, global warming is a great threat facing humanity and it will be the poorest people in the world that suffer most.

Our blue planet; how will humanity alter our home in the twenty-first century?

The Greenhouse Effect

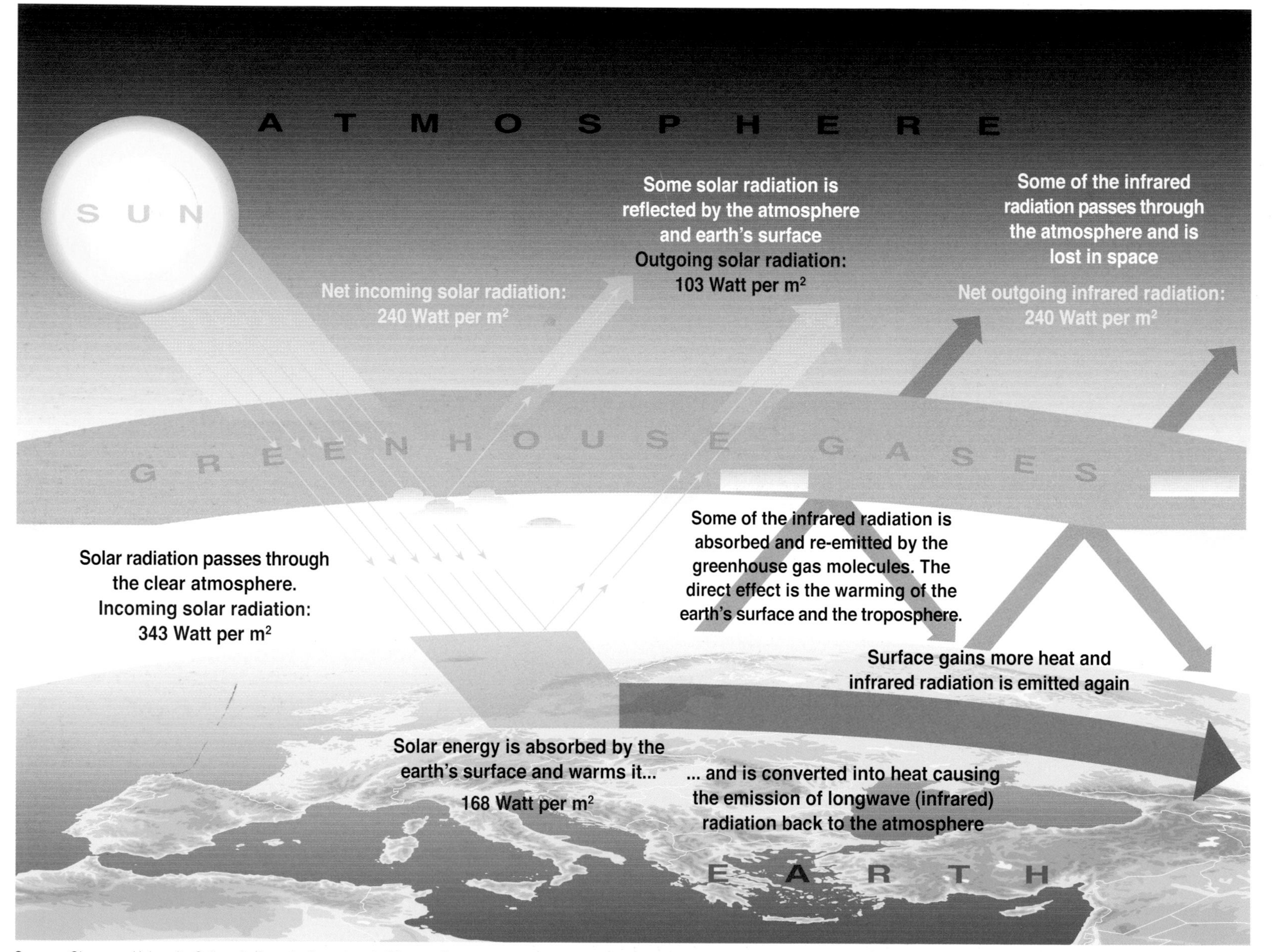

Sources: Okanagan University College in Canada, Department of Geography; University of Oxford, School of Geography; United States Environmental Protection Agency (EPA), Washington; Climate change 1995, The science of climate change, contribution of working group 1 to the second assessment report of the intergovernmental panel on climate change, UNEP and WMO, Cambridge University Press, 1996.

What is Global Warming?

The Earth's Natural Greenhouse

The temperature of the Earth is controlled by the balance of the input of energy of the sun and the loss of this back into space. Certain atmospheric gases are critical to this temperature balance and are known as greenhouse gases. The energy received from the sun is in the form of short-wave energy or 'radiation' (visible and ultra violet light). On average, about one third of the solar radiation that hits the earth is reflected back to space. Of the remainder, most is absorbed by the land and oceans. The Earth's surface becomes warm and as a result emits long-wave 'infrared' radiation. The greenhouse gases can trap some of this long-wave radiation, thus warming the atmosphere. Naturally occurring greenhouse gases include water vapor, carbon dioxide, ozone, methane and nitrous oxide, and together create a natural greenhouse, or blanket effect, warming the Earth by 63°F (35°C).

Although the greenhouse gases are depicted in the figure as a layer, this is only to show their 'blanket effect', as they are in fact mixed throughout the atmosphere.

Human activities are causing greenhouse gas levels in the atmosphere to increase. This increase is causing huge concern as most scientists believe this will lead to an enhanced greenhouse affect, or global warming, the consequences of which will affect everyone on the planet.

Another way to understand the Earth's natural 'greenhouse' is by comparing it with its two nearest neighbors. A planet's climate is decided by its mass, its distance from the sun and the composition of its atmosphere, in particular the amount of greenhouse gases. Mars is smaller than the Earth, so its gravity is less and hence can only retain a smaller atmosphere. Its atmosphere is about a hundred times thinner than Earth's and consists mainly of carbon dioxide. The average surface temperature of Mars is about –50°C, so most of Mars's carbon dioxide is frozen in the ground.

Venus has almost the same mass as Earth but a much thicker atmosphere, which is composed of 96 per cent carbon dioxide. So global warming due to this huge amount of carbon dioxide is intense, producing a surface temperature on Venus of +460°C.

The greenhouse effect: the diagram shows the amount of solar energy or 'radiation' (measured in watts per sq meter) which hits the Earth and how the land, sea and atmosphere absorb or reflect that energy.

In comparison the Earth's atmosphere is very different and this is because of the effects of life on Earth. Our atmosphere is composed of 78 per cent nitrogen, 21 per cent oxygen, and 1 per cent other gases. It is these 'other' gases which we are interested in as they include the so-called greenhouse gases. The two most important greenhouse gases are carbon dioxide and water vapor. Currently carbon dioxide accounts for just 0.03 to 0.04 per cent, while water vapor varies from 0 to 2 per cent. These greenhouse gases are extremely important because they act as a partial blanket for the Earth but at the same time do not overheat our planet. This blanketing is known as the 'natural greenhouse effect'. Without the greenhouse gases, Earth's average temperature would be roughly –4°F (–20°C). The comparison with the climates on Mars and Venus is very stark all because of the different thickness of their atmospheres and the relative amounts of greenhouse gases. However, because of the ease with which the small amounts of greenhouse gases can change on Earth, we know that the Earth's climate is naturally unstable and rather unpredictable compared with that of the other two planets.

Cars are a major source of greenhouse gases.

Past Climate and the Role of Carbon Dioxide

One of the main reasons that we know that carbon dioxide is important in controlling global climate is because of the study of past climate. Over the last two and a half million years the Earth's climate has cycled between the great ice ages, with ice sheets over 1¾ miles (3 km) thick over North America and Europe, to conditions that were even milder than today. These changes are extremely rapid if compared with normal geological variations, which take many millions of years; this is because of the sensitivity of the Earth's climate system. But how do we know about these massive ice ages and the role of carbon dioxide? The evidence comes from ice cores drilled both in Antarctica and Greenland. As snow falls it is light and fluffy and contains a lot of air. When this is slowly compacted to

Planets and Atmospheres

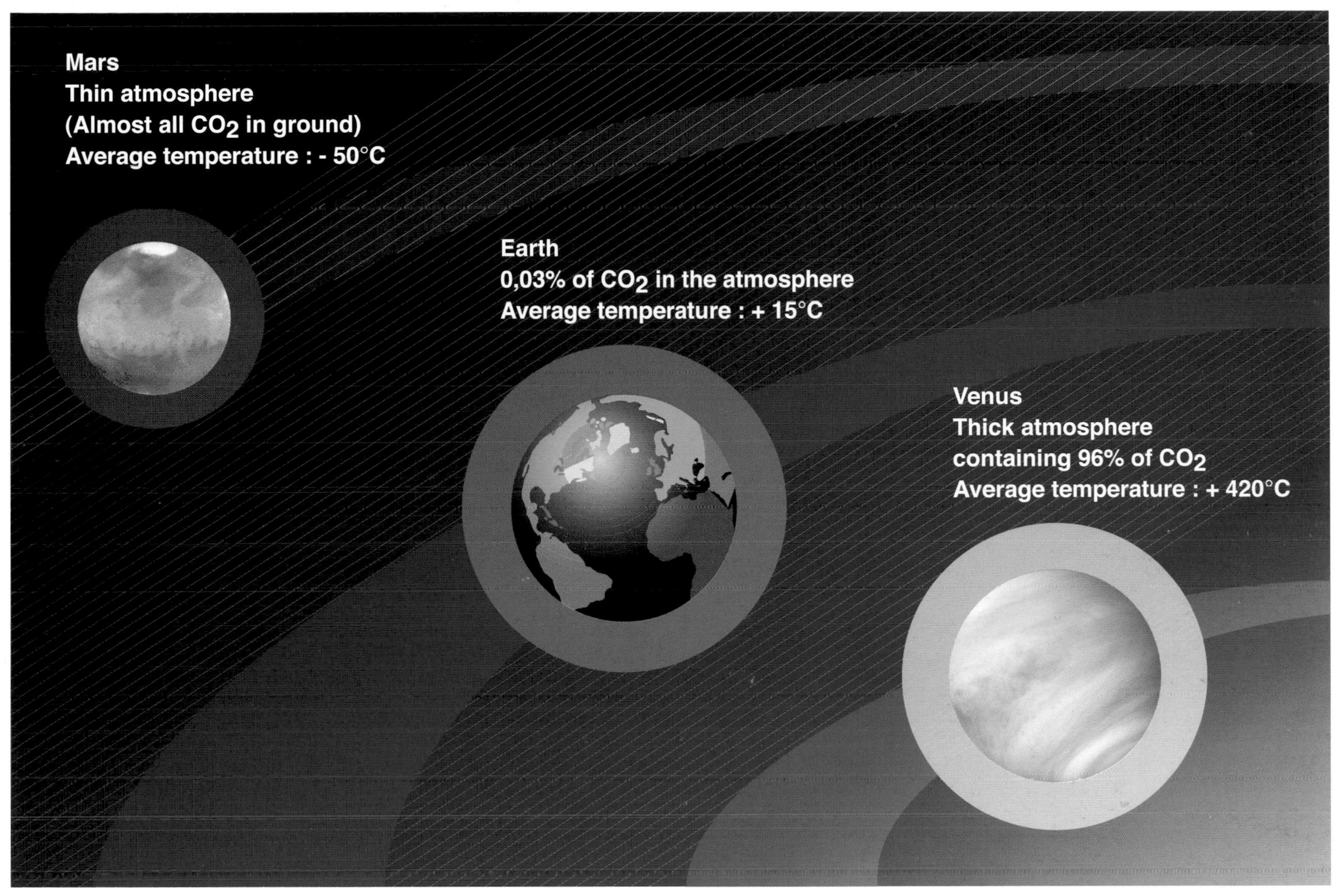

Sources: Calvin J. Hamilton, Views of the solar system, www.planetscapes.com; Bill Arnett, The nine planets, a multimedia tour of the solar system, www.seds.org/billa/tnp/nineplanets.html

The uniqueness of Earth's climate can be seen when it is compared with our closest neighbors, Mars and Venus. The difference in carbon dioxide in the atmosphere produces average temperatures on Mars of –50°C while on Venus they average +420°C.

form ice some of this air is trapped. By extracting these air bubbles from deep within the ancient ice scientists can measure the past content of the atmosphere. It is also possible to measure the temperature at which the ice was formed. The results were amazing, as both atmospheric carbon dioxide and temperatures have covaried over the last 400,000 years. This strongly supports the idea that carbon dioxide content in the atmosphere and global temperature are closely linked, which is our greatest concern for future climate. The study of past climate gives us lots of clues to what could happen in the future. One of the most frightening results from the study of ice cores, lake and deep-sea sediments is that past climate has varied regionally by at least 5.5°F (3°C) in a few decades, suggesting climate can change dramatically on human time-scales. Hence we should expect sudden and dramatic surprises when greenhouse gas levels reach an as-yet-unknown trigger point in the future.

Smog above Santiago, Chile.

The Rise in Atmospheric Carbon Dioxide in the Industrial Period

Should we be worried about global warming? The reason why we should is that there is clear proof that atmospheric carbon dioxide levels have been rising since the beginning of the Industrial Revolution in the eighteenth century. The first measurements of carbon dioxide concentrations in the atmosphere started in 1958 at an altitude of about 13,000 ft (4000 meters) on the peak of Mauna Loa mountain in Hawaii to be remote from local sources of pollution. They have clearly shown that atmospheric concentrations of carbon dioxide have increased every single year since 1958. The average concentration of approximately 316 parts per million by volume (ppmv) in 1958 rose to approximately 369 ppmv in 1998.

This data from Mauna Loa can be combined with the detailed work on ice cores to produce a

Human Effects on Greenhouse Gases

Greenhouse Gas	Chemical Formula	Pre-Industrial Concentrations	1994 Concentrations	Human Source	Global Warming Potential
Carbon dioxide	CO_2	278 ppmv	358 ppmv (30 per cent increase)	Fossil fuel combustion Land use changes Cement production	1
Methane	CH_4	700 ppbv	1721 ppbv (240 per cent increase)	Fossil fuels Rice paddies Waste dumps Livestock	21
Nitrous oxide	N_2O	275 ppbv	311 ppbv (15 per cent increase)	Fertilizer Industrial processes Fossil fuel combustion	310
CFC-12	CCl_2F_2	0	0.503 ppbv	Liquid coolants / foams	6600
HCFC-22	$CHClF_2$	0	0.105 ppbv	Liquid coolants	1350
Perfluoromethane	CF_4	0	0.070 ppbv	Production of aluminium	6500
Sulfur hexa-fluoride	SF_6	0	0.032 ppbv	Dielectric fluid	24,000

The Intergovernmental Panel on Climate Change (IPCC) has identified the main greenhouse gases, where they come from and their warming potential. The warming potential is calculated assuming that carbon dioxide has a potential of one. What is amazing is that there are other greenhouse gases which are much more dangerous than carbon dioxide but which are still in very low concentrations in the atmosphere.

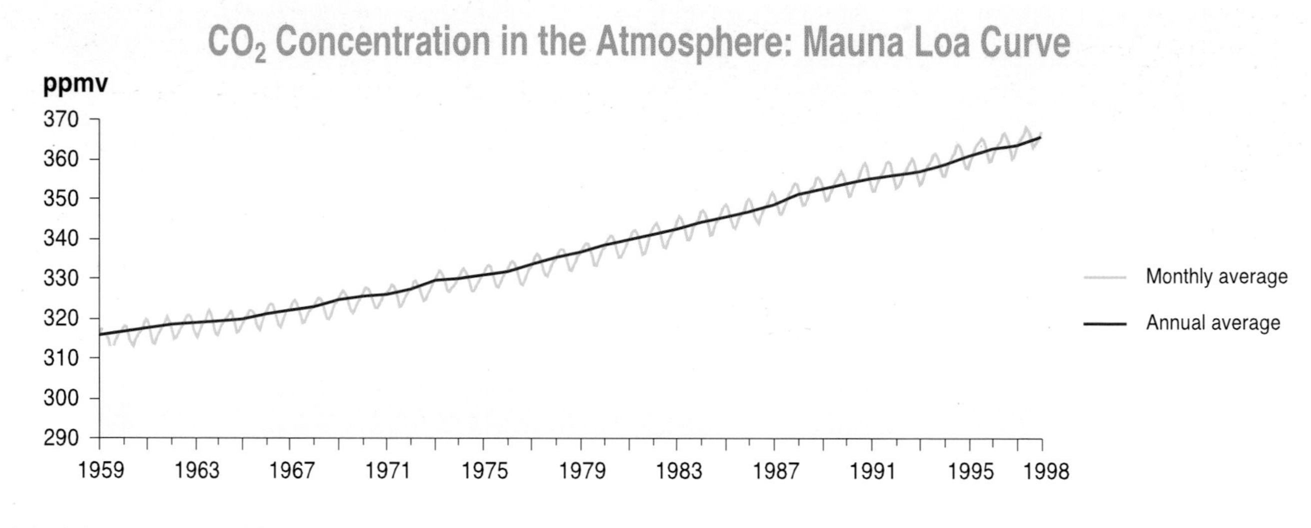

Source : Scripps institution of oceanography (SIO), University of California, 1998.

The concentration of carbon dioxide in the atmosphere has been measured at Mauna Loa in Hawaii. The annual variation seen here is because of carbon dioxide uptake in plants which is greatest during spring in the Northern Hemisphere. This does not, unfortunately, stop the overall trend towards ever-higher values.

complete record of atmospheric carbon dioxide since the beginning of the Industrial Revolution. What this shows is that atmospheric carbon dioxide has increased from a pre-industrial concentration of about 280 ppmv to over 370 ppmv at present, which is an increase of 160 billion tonnes, representing an overall 30 per cent increase.

To put this increase into context we can look at the natural changes between the last ice age, when temperatures were much lower, and the pre-industrial period. According to evidence from ice cores, ice age atmospheric carbon dioxide levels were about 200 ppmv compared to pre-industrial levels of 280 ppm, an increase of over 160 billion tonnes; almost the same carbon dioxide pollution which we have put into the atmosphere. This carbon dioxide increase was accompanied by a global warming of 9°F (5°C) as the world freed itself from the grip of the ice age. Admittedly there are many other causes for the end of the ice age and the subsequent warming, but carbon dioxide did play an important role. What it does demonstrate is that the level of pollution we have caused in 200 years is comparable to natural variations which took thousands of years.

Brief History of Global Warming

Historic Background

Global warming is one of the biggest threats facing humanity and thus it is essential to understand the history of this theory and the evidence that supports it. Global warming has been known about for over a hundred years but was not taken seriously until the late 1980s; I hope I can explain why.

A Swedish scientist, Svante Arrhenius, in pioneering work of 1896, calculated that human activity could, by adding carbon dioxide to the atmosphere, substantially warm the globe. This conclusion was a by-product of his research, the main aim of which was to offer a theory that decreased carbon dioxide might explain the causes of the great ice ages – a theory which still stands today. It was not until 1987 that the Antarctic Vostok ice core results confirmed the pivotal role atmospheric carbon dioxide played in controlling past global climate. However, no one else took up the topic so Arrhenius turned to other challenges. This was because scientists of that period felt there were so many other influences on global climate, from sun spots to ocean circulation, that minor influences by humanity were insignificant compared to the mighty forces of astronomy and geology. This idea was reinforced by classic studies during the 1940s, such as development of the theory that changes in the orbit of the Earth around the sun controlled the waxing and waning of the great ice ages. A second line of argument was that there is 50 times more carbon dioxide in the oceans than the atmosphere and it was conjectured that 'the sea acts as a vast equalizer'; i.e., the ocean would mop up our pollution.

This dismissive view took its first blow when in the 1940s there was a significant improvement in infrared spectroscopy, the technique used to measure long-wave radiation. Up until the 1940s experiments had shown that carbon dioxide blocked the transmission of infrared long-wave radiation of the sort given off by the Earth. However the experiments showed there was very little change in this interception if the amount of carbon dioxide was doubled or halved. That meant even small amounts of carbon dioxide could block radiation so thoroughly that adding more gas made very little difference. Moreover, water vapor, which is much more abundant than carbon dioxide, was found to block radiation in the same way and thus was thought to be more important.

In the original experiments the pressure of the atmosphere at sea level was used. However as you go higher in the atmosphere, the weight of air above gets less and so the pressure drops. This reduction in pressure means that gas molecules are further apart which allows radiation 'energy' to pass through. So increased amounts of carbon dioxide high up in the atmosphere can absorb more radiation. In addition, the argument that water vapor was more important than carbon dioxide was negated when it was discovered that the upper atmosphere was bone dry. This work was brought together by the calculations of Gilbert Plass in 1955 who concluded that adding more carbon dioxide to the atmosphere would mean the atmosphere could intercept more infrared radiation, preventing it being lost to space and thus warming the planet.

This still left the argument that the oceans would soak up the extra anthropogenic, or human-produced, carbon dioxide. The first new evidence came in the 1950s and showed that the average lifetime of a carbon dioxide molecule in the atmosphere before it dissolved in the sea was of the order of ten years. As the ocean over-turning takes several hundreds of years it was assumed the extra carbon dioxide would be safely locked in the oceans. But Roger Revelle, Director of Scripps Institute of Oceanography in California in the 1950s, realized that it was necessary not only to know that a carbon dioxide molecule was absorbed after ten years but to ask what happened to it after that: did it stay there or did it diffuse back into the atmosphere? How much extra carbon dioxide could the oceans hold? Revelle's calculations showed that the complexities of the surface ocean chemistry were such that it returns much of the carbon dioxide that it absorbs relatively quickly. This was the great revelation, that because of the peculiarities of ocean chemistry the oceans would not be a complete sink for anthropogenic carbon dioxide. This principle still holds true, although the exact amount of anthropogenic carbon dioxide taken up per year by the oceans is still in debate; it is thought to be about 2 gigatons, nearly a third of the total annual anthropogenic production.

Charles Keeling, who worked for Roger Revelle, produced the next important step forward in the global warming debate. In the late 1950s and early 1960s Keeling used the most modern technology available to measure the concentration of atmospheric carbon dioxide in Antarctica and Mauna Loa. The resulting Keeling carbon dioxide curves have continued to climb ominously each year since the first measurement in 1958 and have become one of the major icons of global warming.

All the scientific facts about increased atmospheric carbon dioxide and potential global warming were assembled by the late 1950s and early 1960s. In fact Gilbert Plass published an article in 1959 in *Scientific American* declaring that the world's temperature would rise by 5.4°F (3°C) by the end of the twentieth century. The magazine editors published an accompanying photograph of coal smoke belching from factories and the caption read, 'Man upsets the balance of natural processes by adding billions of tons of carbon dioxide to the atmosphere each year'. This resembles thousands of magazine articles, television news items and documentaries which we have all seen since the late 1980s. So why was there a delay between the science of global warming being accepted and in place in the late 1950s and the sudden realization of the true threat of global warming in the late 1980s?

The Delay in Recognizing Global Warming

One of the effects of global warming is an increase in storms.

The key reason for the delay in recognizing the global warming threat is the power of the global average temperature data set (see graph on page 20). This is calculated using the land-air and sea-surface temperature. From 1940 until the mid 1970s the global temperature curve seems to have had a general downward trend, which provoked many scientists to discuss whether the Earth was entering the next great ice age. This fear developed in part due to increased awareness in the 1970s of how variable global climate had been in the past. The emerging subject of palaeoceanography – the study of past oceans – demonstrated from deep-sea sediments that there were at least 32 glacial-interglacial (cold-warm) cycles in the last two and a half million years, not four as had been previously assumed. But the time-resolution of the early data was about one data point every two thousand years, which was so low that there was no possibility of estimating how quickly the ice ages came and went, only how regularly. It led many scientists and the media to ignore the science revelations of the 1950s and 1960s in favor of global cooling.

It was not until the early 1980s, that the 'new ice age' scenario was questioned. By the late 1980s

Global Average Surface Temperature

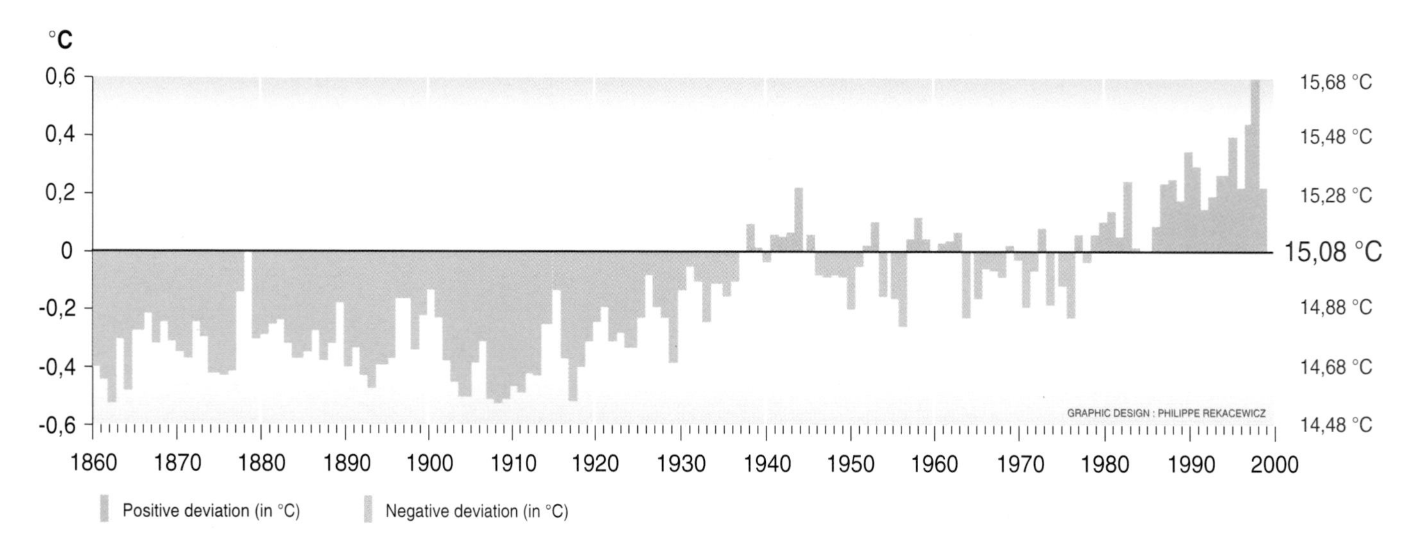

Source: School of environmental sciences, climatic research unit, University of EastAnglia, Norwich, United Kingdom, 1999.

This data set has been constructed using land-air and sea-surface temperatures collected over the last 140 years. The slight cooling in the 1970s can be seen along with the dramatic rise in temperatures over the last 20 years.

the global annual mean temperature curve rose steeply and all the dormant evidence from the late 1950s and 1960s was given prominence; the global warming theory was in full swing.

It seems that the whole global warming issue was driven by the upturn in the global annual average temperature data set, which is surprising because in the 1980s some scientists still believed this was a flawed data set, as: 1) many of the land monitoring stations have since been surrounded by urban areas, thus increasing the temperature records due to the urban heat island effect; 2) there have been changes in the ways ships measure the seawater temperature; 3) it is not possible to tell whether the 1970s cooling and the subsequent warming trend is part of some natural variation or caused by human activities. The IPCC 2001 Science Report, however, uses a wide range of data sets to show that, essentially, the trend in the temperature data is correct, and that this warming trend has continued until the present day. In fact we know that 1997 and 1998 were the two hottest years in the world on record.

The upturn in the global annual average temperature data was not the sole reason for the appearance of the global warming issue. During the 1980s there was also an intense drive to understand past climate change. The major advances were made in obtaining past climate records from deep-sea sediments and ice cores as the time-resolution was increased to a data point at least for every past decade, sometimes years. It was thus realized that glacial periods or ice ages take tens of thousands of years to occur, primarily because ice sheets are very slow to build up and are naturally unstable. In contrast, the transition to warmer periods or interglacials such as the present is geologically very quick, in the order of a couple of thousand years. This is because once the ice sheets start to melt there are a number of processes that can accelerate it, such as sea level rise which can undercut and destroy large ice sheets. The realization that global warming is much easier and more rapid than cooling also put to rest the myth of the next impending ice age. As the glacial-interglacial periods of the last two and a half million years have been shown to be forced by the changes in the orbit of the Earth around the sun, it is possible to predict when the next glacial period would begin if there were no anthropogenic effects: we are not due for another ice age to begin for at least 5000 years. Moreover, recent work on the ice cores and deep-sea sediments demonstrates that at least regional climate changes of a couple of degrees Celsius can occur in a matter of decades. The global climate system is not benign but highly dynamic and prone to rapid changes.

The next change that occurred during the 1980s was a massive grass-roots expansion in the environmental movement, particularly in the USA, Canada, and the UK, partly as a backlash against the right-wing governments of the 1980s and the expansion of the consumer economy and partly because of the increasing number of environmental-related stories in the media. This heralded a new era of global environmental awareness and transnational NGOs (Non-Governmental Organisations). The roots of this growing environmental awareness can be traced back to a number of key markers; these include the publication of Rachel Carson's *Silent Spring* in 1962, the image of Earth seen from the moon in 1969, the Club of Rome's 1972 report, *Limits to Growth*, the Three Mile Island nuclear reactor accident in 1979, the nuclear accident at Chernobyl in 1986, and the *Exxon Valdez* oil spillage in 1989. But these environmental problems were all regional in effect, i.e. limited geographically to the specific area in which they occurred.

It was the discovery in 1985 by the British Antarctic Survey of depletion of ozone over Antarctica which demonstrated the global connectivity of our environment. The ozone 'hole' also had a tangible international cause, the use of CFCs, which led to a whole new area of politics, the international management of the environment. There followed a set of key agreements, the 1985 Vienna Convention for the Protection of the Ozone Layer, the 1987 Montreal Protocol on Substances that Deplete the Ozone layer, and the 1990 London and 1992 Copenhagen Adjustments and Amendments to the Protocol.

The other reason for the acceptance of the global warming hypothesis was the intense media interest throughout the late 1980s and 1990s. This is because the global warming hypothesis was perfect for the media: a dramatic story about the end of the world as we know it, with important controversy about whether it was even true. It was through this media filter that scientists attempted to advance their particular global warming view, either by making claims for more research or promoting certain political options. Scientists became very adept from the late 1980s onwards at staging their media performances, and it is clear that the general acceptance of the global warming hypothesis is in part due to their continued effort to communicate their findings.

If every family in India and China owned a car there would be an extra billion cars in the world.

So combining: 1) the science of global warming, 2) the frightening upturn in the global temperature data set, 3) our increased knowledge of how past climate has reacted to changes in atmosphere carbon dioxide and 4) the massive explosion of the political environmental movement in the late 1980s, leads to the final recognition of the threat of global warming. But who is producing the pollution, what is the evidence for global warming and what are going to be the effects on the planet?

Ice floes in the Arctic Ocean. The amount of ice floes is a good indicator of whether the climate is warming up or cooling down.

Who is Producing the Pollution?

In July 2001, the leaders of the world met in Bonn, Germany, and produced the first international agreement on reducing global carbon dioxide emissions. However, this agreement has been made without the inclusion of the United States, and with very small reduction targets. Fossil fuel emissions are not evenly distributed around the world because of the unequal distribution of industry. A significant amount of carbon dioxide emission comes from energy production, industrial processes and transport, hence many countries believe the agreement may affect certain countries' economies more than others. The industrialized countries must bear the main responsibility for reducing emissions of carbon dioxide, and North America, Europe and Asia emit over 90 per cent of the global human-produced carbon dioxide.

A second very important source of carbon dioxide is due to land-use changes, mainly the cutting down of forests for agriculture, urbanization or roads. When large areas of rainforests are cut down, the land often turns into less productive grasslands with considerably less capacity for storing carbon dioxide.

Who and what are causing global warming? Historically the rich, industrialized countries of the world have emitted most of the anthropogenic greenhouse gases since the start of the Industrial Revolution in the latter half of the 1700s. A major issue of continued debate is the sharing of responsibility. Non-industrialized countries are striving to increase the standard of living for their populations, thereby also increasing their emissions of greenhouse gases, since economic development is closely associated with energy production. Thus the volume of carbon dioxide will probably increase, despite the efforts to reduce emissions in industrialized countries. For example, China has the second biggest emissions of carbon dioxide in the world; however, the Chinese emissions are ten times lower per capita than in the United States, who are top of the list. So this means in the United States every person is responsible for producing ten times more carbon dioxide pollution than in China.

Deforestation releases carbon dioxide into the atmosphere, contributing to global warming.

The Intergovernmental Panel on Climate Change (IPCC) was established in 1988 jointly by the United Nations Environmental Panel and the World Meteorological Organization. The purpose of the IPCC is the continued assessment of the state of knowledge on the various aspects of climate change, including science, environmental and socio-economic impacts and response strategies. The last IPCC report was published in 2001 and took two years to write with 122 main authors, 515 contributing authors and another 450 scientists who reviewed the evidence before it was published. The IPCC is recognized as the most authoritative scientific and technical voice on climate change, and its assessments have had a profound influence on the negotiators of the United Nations Framework Convention on Climate Change (UNFCCC) and its Kyoto Protocol, within which are laid out the general international agreements to cut carbon dioxide pollution. The meetings in the Hague in November 2000 and in Bonn in July 2001 were the second and third attempts to ratify the Protocols laid out in Kyoto in 1998. Unfortunately, President George W. Bush pulled the United States out of the negotiations in March 2001. However, over 180 other countries made history in July 2001 by agreeing the most far-reaching and comprehensive environmental treaty the world has ever seen. The Kyoto Protocol was fully ratified (made legal) on 16 February 2005. It is hoped that the United States will rejoin the negotiations and sign the treaty in the future. The IPCC also provides governments with scientific, technical and socio-economic information relevant to evaluating the risks and developing a response to global climate change.

The use of fossil fuels is the main cause of global warming.

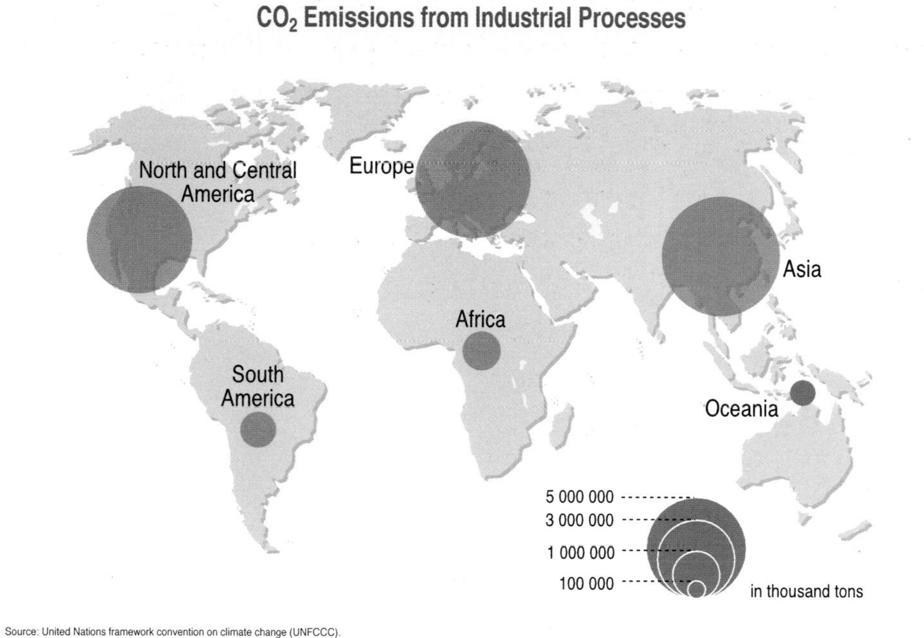

Source: United Nations framework convention on climate change (UNFCCC).

The greenhouse gas carbon dioxide is produced both by industrial processes and land-use changes. These, however, are not evenly distributed around the World. For example North America, Europe and Asia produce 95 per cent of the industrial carbon dioxide.

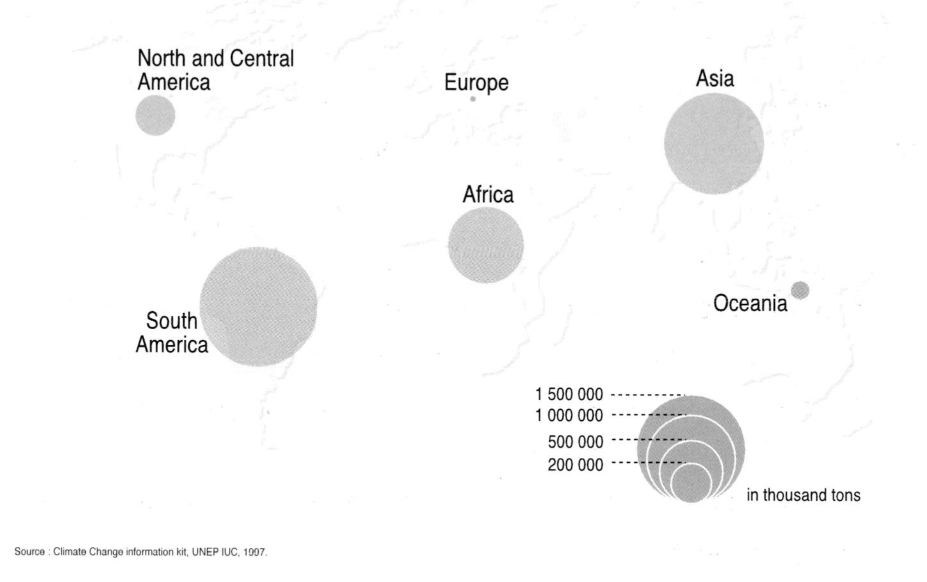

Source : Climate Change information kit, UNEP IUC, 1997.

In contrast 95 per cent of the emissions of carbon dioxide due to changes in land-use are in South America, Africa and Asia. These land-use changes are primarily deforestation for agriculture, urbanization, and roads.

The Evidence for Global Warming

Trends in Global Temperature, Precipitation and Sea Level

The three main indicators of global warming are temperature, precipitation and sea level. The Intergovernmental Panel on Climate Change (IPCC) has combined all the land-surface air and sea surface temperatures (degrees Centigrade) from 1861 to 1998. This data is shown relative to the average temperature between 1961 and 1990, and there has been a sharp warming from the start of the 1980s onwards. The mean global surface temperature has increased by about 0.5° and 1.0°F (0.3° and 0.6°C) since the late nineteenth century and by about 0.3° to 0.5°F (0.2 to 0.3°C) over the last 40 years, which is the period with most reliable data. Recent years have been among the warmest since 1860 – the period for which instrumental records are available. This warming is evident in both sea and land temperatures. In fact we know that 1998 was globally the warmest year on record, with 2002 the second, 2003 the third, 2004 the fourth, 1991 the fifth and 1997 the sixth warmest. The ten warmest years on record have all occurred since 1990.

The warming has not been globally uniform. The recent warming has been greatest between latitudes 40°N and 70°N, though some areas such as the North Atlantic Ocean have cooled in recent decades. Some scientists suggest that part of the warming over the last 30 years is due to increased outputs of energy from the sun. However this is not enough to explain the observed half degree

Sensitive areas such as Hubbard Glacier in Alaska (left) and Kerzas Oasis in the Sahara Desert (above) are providing us with the first warnings that our climate is changing.

warming, and hence we have strong evidence for human-induced global warming.

The IPCC has also spent the last ten years trying to put together all the precipitation records from around the world. Unfortunately, rainfall and snow records are not as well documented as temperature records, and do not go as far back; however, they have been able to assemble general trends in precipitation since the 1900s. It seems that precipitation has increased over land at high latitudes of the Northern Hemisphere, especially during the cold season. But decreases in precipitation occurred in steps after the 1960s over the subtropics and the tropics from Africa to Indonesia. These changes are consistent with available data analyses of changes in stream flow, lake levels and soil surface. Precipitation averaged over the Earth's land surface increased from the start of the century up to about 1960, but has decreased since about 1980. But yet again we have a huge gap in available information, which is due to the lack of data on precipitation over the oceans.

The IPCC has also put together all the available data on sea level and in general it shows that over the last 100 years, the global sea level has risen by about 4 to 10 in (10 to 25 cm). But sea-level change is difficult to measure. Relative sea-level changes have been derived mainly from tide-gauge data. In the conventional tide-gauge system, the sea level is measured relative to a land-based tide-gauge benchmark. The major problem is that the land experiences vertical movements (e.g. from continued recovery from the ice sheets, earthquakes and sedimentation), and these get incorporated into the measurements. With improved methods of measuring the effects of long-term vertical land movements and a better understanding of tide-gauge records, scientists are confident that the volume of ocean water has been increasing.

It is likely that 2 to 4 in (4 to 10 cm) of the rise in sea level over the last 100 years has been related to the concurrent rise in global temperature. On this time scale, the warming and the consequent thermal expansion of the oceans may account for about 1 to 3 in (2 to 7 cm) of the observed sea level rise, while the observed retreat of glaciers and ice caps may account for about 1 to 2 in (2 to 5 cm). Other factors are more difficult to quantify. The rate of observed sea-level rise suggests that there has been some melting of the huge ice sheets of Greenland and Antarctica, but observations of the ice sheets do not yet allow meaningful quantitative estimates of their separate contributions. The ice sheets remain a major source of uncertainty in accounting for past changes in sea level because of insufficient data about them over the last 100 years.

Storms and related floods over the last 50 years account for half the fatalities and economic losses and three quarters of the insured losses from all natural catastrophes. Some predictions suggest that global warming will cause the World to become a more stormy place, for example increasing the number of hurricanes and cyclones by 50 per cent over the next 50 years. This will produce more devastation of both the natural environment and of human settlements as illustrated by the damage caused by Cyclone Hugo (above) and Hurricane Mitch (right).

Melting of the Ice Caps

One of the biggest unknowns of global warming is whether the massive ice sheets over Greenland and Antarctica will melt. One of the key indicators of the expansion or contraction of these ice sheets is the extent of the sea ice which surrounds them. The state of the cryosphere (the global ice) is extremely important as shrinkage of ice cover causes the sea level to rise.

Sea-ice draft is the thickness of the part of the ice that is submerged under the sea. Hence to understand the effects of global warming on the cryosphere it is important to measure how much ice is melting in the polar regions. Comparison of sea-ice draft data acquired on submarine cruises between 1993 and 1997 with similar data acquired between 1958 and 1976 indicates that the average ice draft at the end of the melt season has decreased by about 4 ft (1.3 m) in most of the deep water portion of the Arctic Ocean, from 10 ft (3.1 m) in 1958-1976 to 6 ft (1.8 m) in the 1990s. In summary: ice draft in the 1990s is over 3 ft (1 m) thinner than two to four decades earlier. The main draft has decreased from over 10 ft (3 m) to less than 6 ft 6 in (2 m), and the volume is down by some 40 per cent. In addition in 2000, for the first time in recorded history, a hole large enough to be seen from space opened in the sea ice above the North Pole – another piece of evidence to suggest the world is getting warmer.

Other evidence for global warming comes from the high latitude and high altitude areas where it is so cold that the ground is frozen solid to a great depth. This frozen ground is called permafrost and during the summer months only the top 3 ft (1 m) or so gets warm enough to melt, and this is called the active layer. Already in Alaska, there seems to have been a 5.4°C (3°C) warming down to at least 3 ft (1 m) over the last 50 years, showing that the active layer has become deeper. With the massive increases in atmospheric carbon dioxide predicted for the future, it is likely that there will be increases in the thickness of the active layer of permafrost, or in some areas the complete disappearance of permafrost, over the next century. This widespread loss of permafrost will produce a huge range of problems in local areas as it will trigger erosion or subsidence, change hydrologic processes, and even more carbon dioxide and methane trapped as organic matter in the frozen layers will be released into the atmosphere. Changes in permafrost will reduce the stability of slopes and thus increase incidence of slides and avalanches. A more dynamic cryosphere will increase the natural hazards for people, structures and communication links. Buildings, roads, pipelines, such as the oil pipelines, and communication links are already being threatened in Alaska and Siberia.

Changing Weather

There is lots of evidence that our weather patterns are changing. In recent years massive storms and subsequent floods have hit China, Italy, England, Korea, Bangladesh, Venezuela, India and Mozambique. The winter of 2000/2001 was the wettest six months recorded in Britain since records began in the eighteenth century. In addition, on average, British birds nest two weeks earlier than 30 years ago. Insect species – including bees and termites – that need warm weather to survive are moving northward, and some have already reached England from France. Glaciers in Europe are in retreat, particularly in the Alps and Iceland. Ice-cover records from the Tornio River in Finland show that the spring thaw of the frozen river now occurs a month earlier than when records began in 1693.

There is also evidence for more storms in the North Atlantic. Wave height in the North Atlantic Ocean has been monitored since the early 1950s, from light ships, ocean weather stations and, more recently, satellites. Between the 1950s and 1990s the average wave height has increased from 8 ft to 11 ft 6 in (2.5 to 3.5 m), an increase of 40 per cent. Storm intensity is the major control on wave height so there is clear evidence for a major surge in storm activity over the last 40 years.

Concerning hurricanes in the Atlantic Ocean, there is at the moment a huge debate among scientists whether there is evidence of an increase in the number of hurricanes over the last 100 years. With 6 large hurricanes hitting Florida in the 2005/6 season and Hurricane Katrina destroying New Orleans, the evidence seems to suggest there is an upturn in the number of hurricanes. Combine this with the fact that in 2004 Hurricane Catarina became the first hurricane in recorded history to appear in the South Atlantic Ocean, wreaking havoc along the Brazilian coast.

In addition, property losses in the United States alone have increased tenfold in the last 30 years, mainly because of the increased concentration of valuable property and infrastructure in low-lying coastal regions. But with increasing global warming, it will be easier to achieve the critical temperatures in the oceans required to form hurricanes, spawning more of them with more energy to unleash upon our coastlines. The message is clear, the Caribbean and United States in our 'greenhouse world' could be hit more often and by bigger, meaner hurricanes.

Global warming means more hurricanes in the future with all the destruction they bring.

El Niño *events occur in the Pacific Ocean; however, their influence is world-wide and global warming may make them more common.* El Niño *events can influence the direction of hurricanes and cyclones as well as causing droughts in the southern United States, East Africa, northern India, northeast Brazil and Australia. Torrential rains and terrible floods can occur in California, South America, Sri Lanka and east central Africa. An example of the latter were the massive floods that hit Africa, as illustrated by the Tana River flooding a village near Garsen during the* El Niño *of 1998 (above left).*

Climate change is a problem for everyone in the world. Even in the relatively mild climate of Europe changes have already started. Massive flooding occurred throughout Europe in both in 2000 and 2001; particularly severely hit were Britain and Italy. These extreme events, such as the flooding of Venice (below left), will become more commonplace in the future.

Increased Frequency of *El Niño*

One of the most important and mysterious elements in global climate is the periodic switching of direction and intensity of ocean currents and winds in the Pacific Ocean. Originally known as *El Niño* ('Christ child' in Spanish), as it usually appears at Christmas, and now more normally known as ENSO (El Niño-Southern Oscillation), this phenomenon typically occurs every three to seven years. It may last from several months to more than a year. The 1997-8 *El Niño* conditions were the strongest on record and caused droughts in southern United States, East Africa, northern India, northeast Brazil and Australia. In Indonesia, forest fires burned out of control in the very dry conditions. In California, parts of South America, Sri Lanka and east central Africa there were torrential rains and terrible floods.

ENSO is an oscillation between two climates, 'Normal' and '*El Niño*' conditions. *El Niño* conditions have been linked to changes in the monsoon, storm patterns and occurrence of droughts all over the world. The state of the ENSO has also been linked to the position and occurrence of hurricanes in the Atlantic Ocean. It is thought that the poor prediction of where Hurricane Mitch made landfall in 1998 was because the ENSO conditions were not considered.

There are also fears that ENSO has been affected by global warming. The *El Niño* conditions should only occur every three to seven years, but they returned for three years out of four, 1991-1992, 1993-1994, 1994-1995, 1997-1998 and again in 2002-2003, wreaking havoc on global weather. Scientists at the University of Colorado have drilled a core from the coral reefs in the Western Pacific which record sea surface temperature back 150 years, well beyond historic records. The sea surface temperature shows the shifts in ocean current which accompany shifts in the ENSO. The core evidence shows that there have been two major changes in the frequency and intensity of *El Niño* events. The first change was at the beginning of the twentieth century, from a 10 to 15 year cycle to a three to five year cycle. The second shift was a sharp threshold in 1976 when a marked change to more intense and more frequent *El Niño* events occurred.

These are sobering results considering the huge weather disruption and disasters caused by the last few *El Niño* years. Modeling results also suggest that the current 'heightened' state of *El Niño* can permanently shift weather patterns. For example, it seems that the drought region in the United States could be shifting eastward.

The Present Carbon Cycle

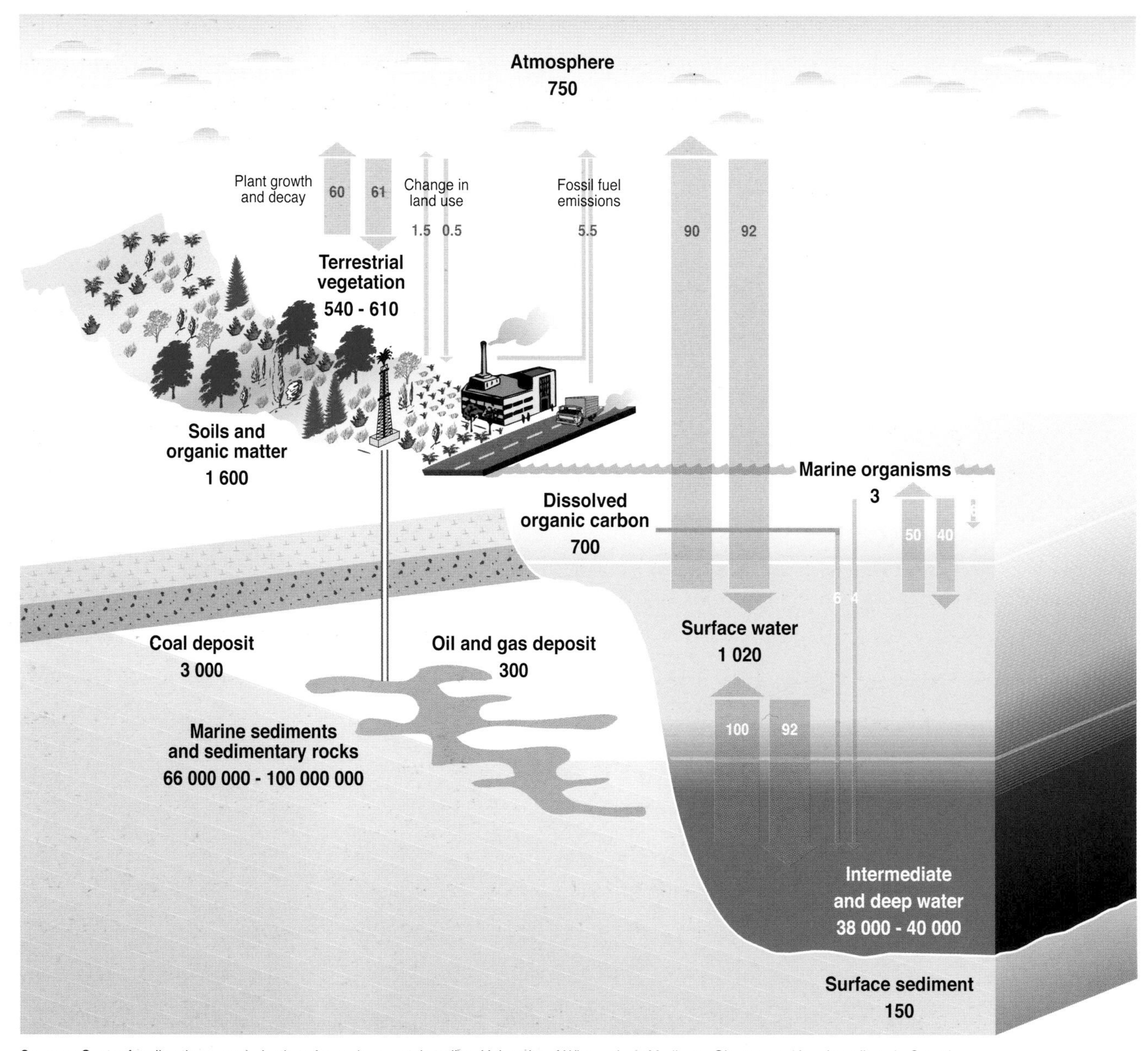

Sources: Center for climatic research, Institute for environmental studies, University of Wisconsin at Madison; Okanagan university college in Canada, Department of geography; World Watch, November-December 1998; Climate change 1995, The science of climate change, contribution of working group 1 to the second assessment report of the intergovernmental panel on climate change, UNEP and WMO, Cambridge University Press, 1996.

The Impacts of Global Warming

Humanity's greenhouse gas emissions, as we have seen, have already started to change our climate and are expected to lead to major environmental and societal changes in the twenty-first century and beyond. These changes will potentially have wide-ranging effects on the natural environment as well as on human societies and economies. Scientists have made estimates of the potential direct impacts on various parts of our society, but in reality the full consequences will be more complicated because of the complexity of the world's political and economic systems.

To assess potential impacts, it is necessary to estimate the extent and magnitude of climate change, especially at national and local levels. The IPCC has recently published its assessment of climate change on the United States, which deals with the impacts on a region by region basis. Although much progress has been made in understanding the climate system and climate change, it must be remembered that projections of climate change and its impacts still contain many uncertainties, particularly at regional and local levels.

Why is it So Difficult to Model the Future?

To be able to predict the effects of global warming we need to understand the present-day carbon cycle. The Earth's carbon cycle is extremely complicated with both sources and sinks of carbon dioxide. The figure opposite shows the global carbon reservoirs in GtC (gigatonne = one thousand million tonnes) and fluxes, i.e., the ins and outs of carbon in GtC/year. These indicated figures are annual averages over the period 1980 to 1989. But it must be remembered that the component cycles are simplified and the figures present only average values. The carbon transported by rivers, particularly the anthropogenic portion, is currently very poorly quantified and is not shown here. Evidence is accumulating that many of the exchanges of carbon can fluctuate significantly from year to year. In contrast to the static view conveyed in figures like this one, the carbon system is dynamic and coupled to the climate system on

The present carbon cycle is complex as not only are there stores of carbon (dark numbers) but there are also exchanges of 'fluxes' of carbon which are continually occurring (light numbers). All the stores and exchanges of carbon are given in billions of tonnes of carbon.

seasonal, interannual and decadal timescales. The most interesting figure is that the surface ocean takes up just less than half the carbon dioxide produced by industry per year. However, this is one of the most poorly known figures and there is still considerable debate about whether the oceans will continue to be such a large sink for our pollution. As we will see later in this book, one of the great surprises recently has been the completely unexpected massive uptake of atmospheric carbon dioxide by the Amazon rainforest. The key question, is how long will this last?

Coal-fired power stations produce both carbon dioxide and aerosols.

An added complication is that there are also cooling effects. These include the amount of aerosols in the air, many of which come from human pollution, such as sulfur emissions from power stations. They have a direct effect on the amount of solar radiation hitting the Earth's surface, as they reflect the sun's energy back into space before it reaches the earth. Aerosols may have significant local or regional impact on temperature. Some scientists have called this effect 'global dimming' as the amount of sunlight hitting the ground diminishes. In fact the United Kingdom Meteorology Office has figured them into their computer simulations of global warming and they provide an explanation of why industrial areas of the planet have not warmed as much as previously predicted. Water vapor is a greenhouse gas, but at the same time the upper white surface of clouds reflects solar radiation back into space. This reflection is called 'albedo'; clouds and ice have a high albedo and hence reflect large quantities of solar radiation from surfaces on the Earth. Predicting what will happen to the amount and types of clouds and the extent of

The Cooling Factors

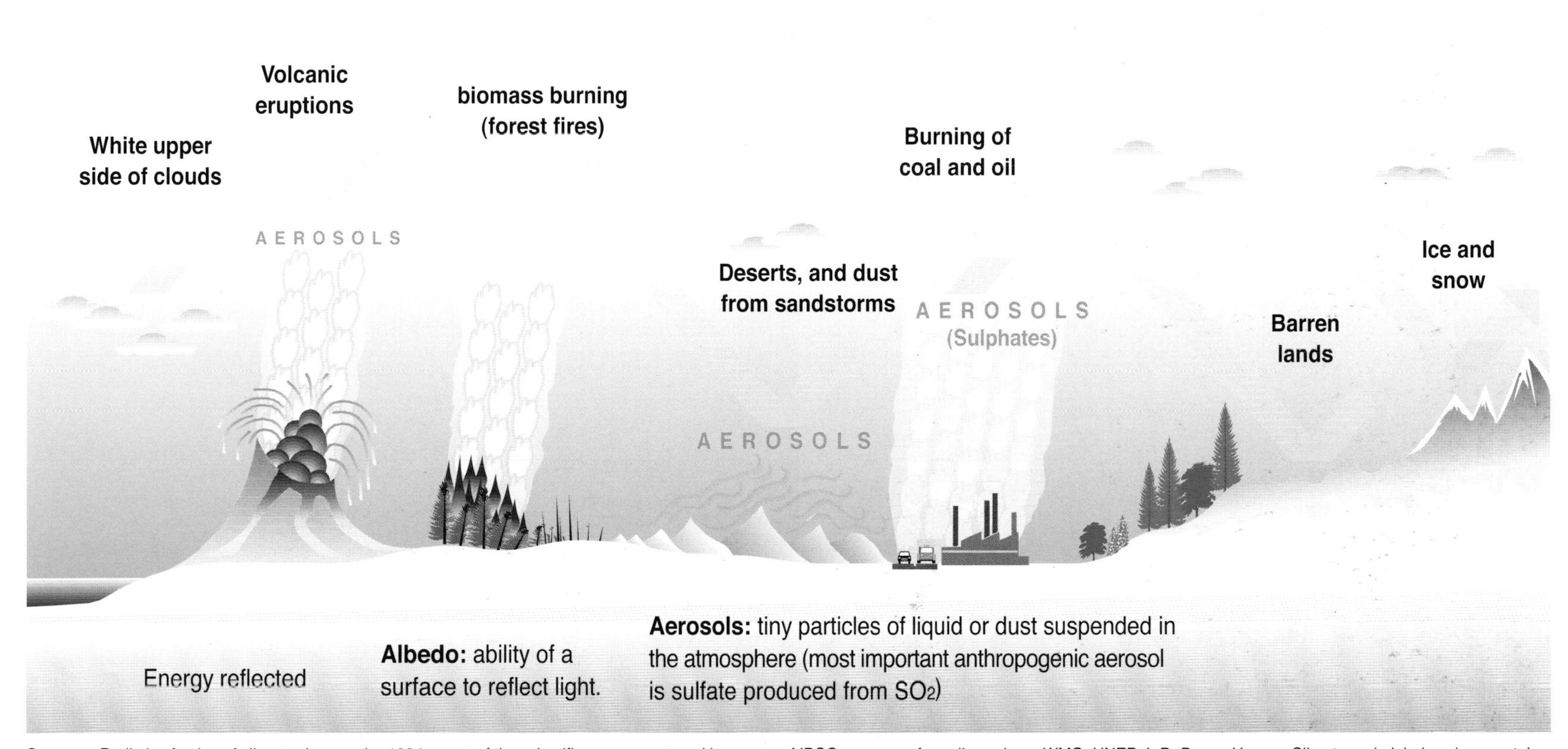

Sources: Radiative forcing of climate change, the 1994 report of the scientific assessment working group of IPCC, summary for policymakers, WMO, UNEP; L.D. Danny Harvey, Climate and global environmental change, Prentice Hall, Pearson Education, Harlow, United Kingdom, 2000.

global ice in the future creates huge difficulties in calculating the exact effect of global warming. For example if the polar icecap melts, the albedo will be significantly reduced. Open water absorbs heat, while white ice and snow reflect it, so this would be a positive feedback, increasing the effects of global warming.

The third problem with trying to predict future climate is predicting the amount of carbon dioxide emissions we will produce in the future. This will be influenced by population growth, economic growth, fossil fuel usage, the rate of deforestation, and whether an international agreement to cut emissions is ever reached. The IPCC has produced a model for the worst– and best–case scenarios for the future. The worst case produces an increase of 220 per cent in atmospheric carbon dioxide compared with pre-industrial levels and the best case produces a 75 per cent increase. Even if the anthropogenic emissions of carbon dioxide are stabilized or reduced, the carbon dioxide content in the atmosphere will still increase for up to 100 years.

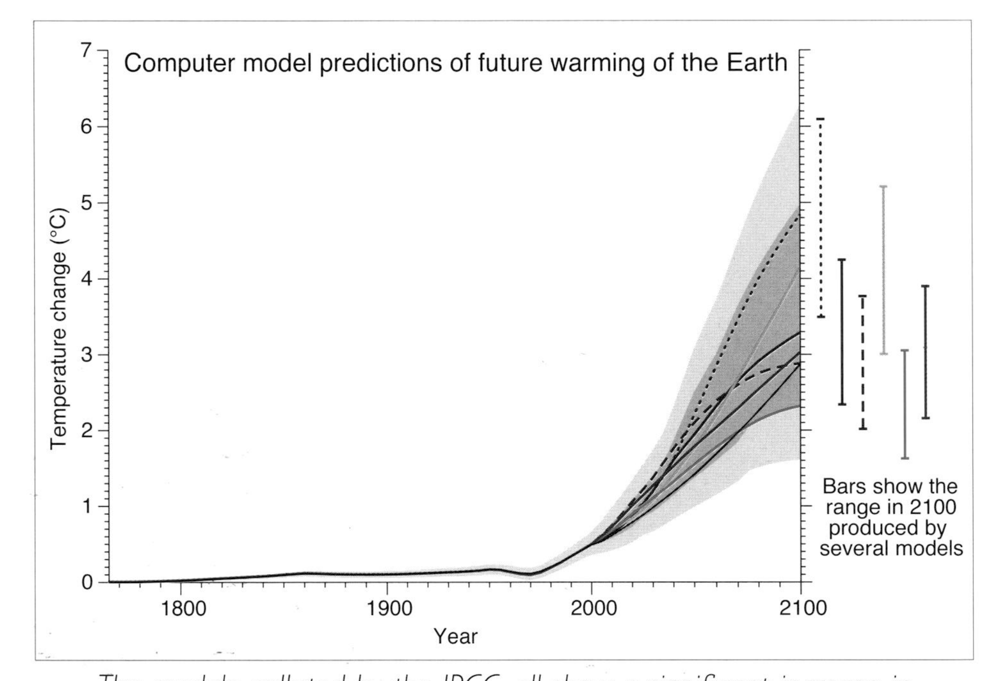

The models, collated by the IPCC, all show a significant increase in temperature by 2100 but with huge uncertainties about exactly how much.

Future Global Temperatures and Sea Level

Using the full range of IPCC 2001 carbon dioxide emission scenarios, the global mean temperature changes relative to 1990 were calculated up to 2100 (see above). These climate models show that the global mean surface temperature could rise by about 2.5° to 10.5°F (1.4° to 5.8°C) by 2100. The topmost curve assumes constant aerosol concentrations beyond 1990 and high climate sensitivity and an increase of 10.5°F (5.8°C) by 2100. The lowest curve assumes constant aerosol concentrations beyond 1990 but a much lower climate sensitivity and an increase of 2.5°F (1.4°C). What is most worrying is that there is still about 6.1°F (3.4°C) temperature difference in the most extreme estimates, which are based on how sensitive scientists believe the climate system to be!

Again using the different carbon dioxide emission scenarios the IPCC has projected global mean sea level up to 2100. Taking into account the ranges in the estimate of climate sensitivity and ice melt parameters, and the emission scenarios, the models project an increase in global mean sea level of between 7¾ in and 35 in (20 and 88 cm). Note that during the first half of this century, the choice of emission scenario has relatively little effect on the projected sea level rise as most of it is due to the large thermal inertia of the ocean-ice-atmosphere climate system. However it has an increasingly large effect in the later part of this century, because of the uncertainty about how the ice sheets will react and melt. In addition, because of the thermal inertia of the oceans, sea level would continue to rise for many centuries beyond 2100 even if concentrations of greenhouse gases were stabilized at that time.

One of the largest uncertainties is whether the Arctic and Antarctic ice will start to melt.

What the sea level calculation does not take into account is the possible melting of the world's ice sheets and glaciers. If all the ice sheets melted their contribution to sea level rise would be as follows: mountain glaciers = 1 ft (0.3 m), Greenland = 23 ft (7 m), Western Antarctic ice sheet

= 28 ft (8.5 m) Eastern Antarctic ice sheet = 213 ft (65 m). What is worrying is that NASA satellite measurements suggest both Greenland and the Western Antarctic ice sheets are shrinking. If this produces enough meltwater then we could have some big surprises in store in the future.

Coastlines

The IPCC predicts that under a business-as-usual-scenario (i.e., continued increase in burning fossil fuels) sea level could rise between 8 and 35 in (20 and 88 cm) in the next 100 years. This is primarily due to the thermal expansion of the oceans, and this is a major concern to all coastal areas as it will decrease the effectiveness of coastal flood defenses and increase the instability of cliffs and beaches. In Britain and the United States the response to this danger of ever-increasing sea levels has been to: add another few feet to the height of sea walls around property on the coast, abandon some poorer quality agricultural land to the sea, as it is no longer worth the expense of protecting it, and add more legal protection to coastal wetlands, being nature's best defense against the sea. However, there are countries, for example small island and river delta based nations, which face a much more dire situation than Britain or the United States.

Flooding in Bangladesh is likely to increase in the future.

For small island nations such as the Maldives in the Indian Ocean and the Marshall Islands in the Pacific Ocean, 3 ft (1 m) rise in sea level would flood up to 75 per cent of the dry land, making the islands uninhabitable. However, there is a different twist to the story if we consider nations with a significant portion of their country on river deltas, such as Bangladesh, Egypt, Nigeria and Thailand. A World Bank report in 1994 concluded that other human activities on the deltas were exacerbating the effects of global warming by causing these delta areas to sink.

In the case of Bangladesh, over three quarters of the country is within the deltaic region formed by the confluence of the Ganges, Brahmaputra and Meghna rivers. Over half the country lies less than 16 ft (5 m) above sea level and flooding is a common occurrence; during the summer monsoon a quarter of the country is flooded. Yet these floods, like those of the Nile, bring with them life as well as destruction. The water irrigates and the silt fertilizes the land. The fertile Bengal Delta supports one of the world's most dense populations, over 110 million people in 54,000 sq miles (140,000 sq km). But the monsoon floods have been getting worse throughout the 1990s. Every year the Bengal Delta should receive over 1 billion tons of sediment and 38 cu. miles (100 cu. km) fresh water. This sediment load balances the erosion of the delta both by natural processes and human activity. However, the Ganges River has been diverted in India into the Hooghly Channel for irrigation. This reduced sediment input is causing the delta to subside. Exacerbating this is the rapid extraction of fresh water from the delta for agriculture and drinking water. In the 1980s, 100,000 tubewells and 20,000 deep wells were sunk, increasing the fresh water extraction sixfold. Both these projects were aimed at increasing the quality of life for people in this region, but have produced a subsidence rate of up to 1 in (2.5 cm) per year, one of the highest rates in the world.

From the worst-case scenario using estimates of subsidence rate and global warming sea level rise, the World Bank has estimated that by the middle of the twenty-first century the relative sea level in Bangladesh could rise by as much as 6 ft (1.8 meters), and they estimated that this would result in a loss of up to 16 per cent of land, supporting 13 per cent of the population and producing 12 per cent of the current gross domestic product (GDP) – money that Bangladesh can ill afford to lose. This scenario does not take any account of the devastation of the mangrove forest and the associated fisheries. Moreover, increased landward intrusions of saltwater would further damage water quality and agriculture. Though Bangladesh is the worst case globally, similar changes are observed at all other major delta regions.

Another example of a threatened coastline is the Nile Delta, which is one of the oldest intensively cultivated areas on earth. It is very heavily populated, with population densities up to 1600 inhabitants per third of a sq mile (1 sq km). Deserts surround the low-lying, fertile floodplains. Only 2.5 per cent of Egypt's land area, the Nile delta and the Nile valley, are suitable for intensive agriculture. Most of a 30 mile (50 km) wide land strip along the coast is less than 6 ft 6 in (2 m) above sea level and is only

The major problem with future climate change is the lack of predictability. All human activities rely on predicting the local weather, nowhere is this more important than in farming. Over the last ten years extreme climate events seem to be becoming more common. In 1998 Hurricane Mitch tore through Central America leaving 20,000 dead. It also destroyed the region's banana plantations (left), crippling the local economy and delaying recovery from the natural disaster.

However, the developed world is not immune to natural disasters and in 2000 flooding in France destroyed some of the country's vineyards (right). If we can make better predictions of future climate change and the likelihood of extreme events then we can adapt and change our farming practices, preventing such disasters.

protected from flooding by a ¾ to 6 mile (1 to 10 km) wide coastal sand belt, shaped by discharge of the Rosetta and Damietta branches of the Nile. Erosion of the protective sand belt is a serious problem and has accelerated since the construction of the Aswan Dam in the south of Egypt.

Rising sea level would destroy weak parts of the sand belt, which are essential for the protection of lagoons and the low-lying reclaimed lands. The impacts would be very serious. About one third of Egypt's fish catches are made in the lagoons. Sea level rise would change the water quality and affect most fresh-water fish. Valuable agricultural land would be inundated. Vital, low-lying installations in Alexandria and Port Said would be threatened. Recreational beach facilities, important for tourism, would be endangered and essential groundwater would be salinated. Dykes and protective measures would probably prevent the worst flooding up to a 20 in (50 cm) sea level rise, but it would cause serious groundwater salination and the impact of increasing wave action would be severe.

Future Storms and Floods

We know from historic records that during periods of rapid climate change, weather patterns become erratic and the number of storms increases. One example of this is the Little Ice Age, which lasted from the end of the sixteenth century to the beginning of the eighteenth century, and is mainly remembered for the ice fairs that were held on the frozen River Thames in London. However, what is not remembered is that going into and coming out of the Little Ice Age there were some apocalyptic tempests. For example, as the climate was finally warming in 1703 at the end of the Little Ice Age, there was the worst recorded storm in British history, which killed over 8000 people.

Storms and floods between 1951 and 1999 were responsible for 76 per cent of the global insured losses, 58 per cent of the economic losses and 52 per cent of the fatalities from natural catastrophes. There is good evidence that the world is getting stormier, resulting in an increase in economic loss and fatalities, especially as global warming is bad news when it comes to the number and intensity of hurricanes. To give you an idea of the destructive capability of hurricanes, in August 2005 Hurricane Katrina hit the United States, devastated the city of New Orleans, killed over 1500 people and caused record damage, estimated at $25 billion. Hurricanes and their cousins cyclones and typhoons form in the tropics when the sea surface temperature is at least 78°F (26°C) down to 200 ft (60 m) below the surface. All it then takes is a further increase of 1.8°F (1°C) in sea surface temperature to reduce atmospheric pressure enough to start the convective cell, a column of rapidly rising air, which sucks in air at sea level and produces the powerful hurricane vortex. With increasing global warming, achieving the critical temperatures in the oceans will be easier than ever before, spawning more hurricanes with more energy to be unleashed upon our coastlines.

The message is clear, the Caribbean, United States, and Central America in our 'greenhouse world' will be hit more often and by bigger, more violent hurricanes. The additional problem is that the direction of these hurricanes is becoming more difficult to predict. The forecast of Hurricane Gilbert in 1988 was extremely accurate so when it hit Jamaica the evacuation procedures were so effective that only 44 people died. This is compared to Hurricane Mitch in 1998, which was the same size and power, and which was assumed to be going northwards; however, it shocked scientists and instead turned west and hit Central America, with the loss of over 20,000 lives.

In comparison, the predictions of Hurricane Katrina in 2005 were very good, but the protection for New Orleans had been neglected, with the levees only high enough for a category 3 hurricane and important wetlands, which act as a buffer to the 15 foot high storm surge, drained. In addition, the evacuation plans were completely inadequate. So in the future there needs to be an acceptance that coastal areas will be increasingly vulnerable and that we need to adapt to the stormier future, with better defences and better evacuation plans all around the world.

Hence it may not be possible to build on prime locations such as the Florida coast, which should be allowed to return to wetland, the best natural defence against the rage of a hurricane.

Biodiversity

Global warming will have a significant effect on the natural world and of particular worry is its effect on biodiversity, as the location of each vegetation type is controlled by the temperature and precipitation. A good example of how global warming will affect plants is the impact of a warmer climate on forests. Deciduous forests will probably move northwards and to higher altitudes, replacing coniferous forests in many areas. However, if climate change is too rapid then tree migration, which takes decades, will not be able to keep up. This will be compounded by the fact that many species will want to migrate into areas which are dominated by urban sprawl and farming. Climate and vegetation changes could occur also in mountainous areas. Mountains cover about 20 per cent of the Earth's continents and serve as an important water source for most major rivers. Records of vegetation and climate have indicated that during past periods of climate warming vegetation zones have shifted to higher elevations, resulting in the loss of some species and ecosystems. Simulated scenarios for temperate-climate mountain sites suggest that continued warming could have similar consequences. Species and ecosystems with limited climatic ranges could disappear and, in most mountain regions, the extent and volume of glaciers and the extent of permafrost and seasonal snow cover will be reduced. Associated changes in precipitation would affect soil stability. Both of these would dramatically affect socio-economic activities from skiing holidays to agriculture, hydropower to logging.

The unique species of mountain cloud rainforests are threatened by future climate change.

The IPCC report lists the following species as those most at threat from climate change predicted due to global warming: the mountain gorilla in Africa, amphibians that only live in the cloud forest of the neotropics, the spectacaled bear of the Andes, forest birds of Tanzania, the Resplendent Quetzal in Central America, the Bengal tiger and other species only found in the Sundarban wetlands, rainfall sensitive plants found only in the Cape Floral Kingdom of South Africa, polar bears and penguins. Natural habitats which are threatened include coral reefs, mangroves, other coastal wetlands, montane ecosystems found in the upper 200-300 m of mountainous areas, prairie wetlands, permafrost ecosystems and ice edge ecosystems which provide a habitat for polar bears and penguins.

One example of an ecosystem under threat is the coral reefs. Coral reefs are a valuable economic resources for fisheries, recreation, tourism and coastal protection. In addition, reefs are one of the largest global stores of marine biodiversity, with untapped genetic resources. Some estimate the global cost of losing the coral reefs runs in to hundreds of billions of dollars each year.

The last few years have seen unprecedented declines in the health of coral reefs. In 1998 *El Niño* was associated with record sea-surface temperatures and associated coral bleaching, which is when the coral expel the algae that live within them and that are necessary to their survival. In some regions as much as 70 per cent of the coral may have died in a single season.

Agriculture

One of the major worries concerning future climate change is the effect it will have on agriculture both globally and regionally. The main question is, can the world feed itself under the predicted future global warming conditions? Predictions of cereals production for 2060 suggests that there are still huge uncertainties about whether climate change will cause global agricultural production to increase or decrease. The models suggest that on a world-wide scale that change is expected to be small or moderate. This masks the huge changes that will occur in different regions, with both winners and losers, the poorest countries, which are least able to adapt, being the losers.

The results of these studies are heavily dependent on the assumed trade models and market forces used, as agricultural production has very little to do with feeding the world's population and much more to do with trade and economics. This is why the EU has stockpiles of food, while many under-developed countries export cash crops (e.g., sugar, coca, coffee, tea, and rubber) but cannot adequately feed their populations. A classic example is the West African state of Benin, where cotton farmers can obtain cotton yields of four to eight times per hectare greater than their US competitors in Texas. The USA subsidizes their farmers, however, which means that US cotton is cheaper than that coming from Benin. Currently, US cotton farmers receive $3.9 billion in subsidies, almost twice the total GDP of Benin. So even if global

The last ten years have seen some of the warmest years ever recorded. These have resulted in exceptionally warm surface waters in our tropical oceans. Unfortunately when the surface of the ocean gets too warm the corals, which live within a few feet of the surface, die. This is called bleaching and has a dramatic effect as it not only kills the corals but also all the other abundant life that they support, as can be seen from the effect of bleached coral on the surrounding waters in the Maldives (above). Compare this with the huge biodiversity that surrounds a heathy coral reef at Layang Layang, Malaysia (right).

In many parts of the world collecting fresh water is the most important job of the day. Predictions of future climate change suggest that water, the most essential resource for life, will become more scarce, forcing longer and harder trips to collect enough water to live.

warming makes Texan cotton yields even lower, it still does not change the biased market forces.

The other completely unknown factor is how adaptable a country's agriculture can be. For example the models assume that developing countries' production levels will fall more compared with those of developed countries, because their estimated capability to adapt is less than in developed countries. But this is just another assumption that has no analog in the past.

One example of the real regional problems that global warming could cause is the case of coffee growing in Uganda, where the total area suitable for growing would be dramatically reduced, to less than 10 per cent, with a temperature increase of 3.6°F (2°C). Only higher areas would remain cultivable, the rest would become too hot to grow coffee. This demonstrates the vulnerability of developing countries, whose economies often rely heavily on one or two agricultural products, to the effects of global warming.

Fresh Water

Today, rising human populations, particularly growing concentrations in urban areas, are putting great stress on water resources. The impacts of climate change – including changes in temperature, precipitation and sea levels – are expected to have varying consequences for the availability of fresh water around the world. For example, changes in river runoff will affect the yields of rivers and reservoirs and the recharging of groundwater supplies. An increase in the rate of evaporation will also affect water supplies and contribute to the salinization of irrigated agricultural lands. Rising sea levels may result in saline intrusion into coastal aquifers. Approximately 1.7 billion people, a third of the world's population, live in countries that are water stressed. IPCC predictions suggest that with projected global population increase and climate change, and assuming present consumption patterns, 5 billion people will experience water stress by 2025. Climate change is likely to have the greatest impact in countries with a high ratio of relative use to available supply. Regions with abundant water supplies will get more than they want with a massive increase in flooding. Computer models predict much heavier rains and thus major flood problems for Europe. Paradoxically, countries that currently have little water (e.g. those relying on desalinization) may be relatively unaffected. It is those countries in between, which have no history or infrastructure for dealing with water shortages, which will be the most affected. In central Asia, North Africa and southern Africa there will be even less rainfall and water quality will become increasingly degraded through higher temperatures and pollutant run-off. Thus the terrible droughts in Africa and massive floods in Europe, the United States and southeast Asia will be the norm in the future.

Diseases

The transmission of many infectious diseases is affected by climatic factors. Infective agents and their vector organisms are sensitive to temperature, surface water, humidity, wind, soil moisture, and changes in forest distribution. This applies particularly to vector-borne diseases, which are carried by another organism, like malaria which is carried by mosquitoes. It is projected that climate change and altered weather patterns would affect the range (both altitude and latitude), intensity, and seasonality of many vector-borne and other infectious diseases. In general, increased warmth and moisture due to global warming will enhance transmission of diseases.

We should remember that our capacity to control the diseases will also change. New or improved vaccination can be expected; some vector species can be constrained by use of pesticides, but there are uncertainties and risks here, too: for example, long-term pesticide use breeds resistant strains and kills many predators of pests.

The most important vector-borne disease is malaria, with 500 million people currently infected world-wide, which is about twice the population of the United States. *Plasmodium vivax*, with the Anopheles mosquito as a vector, is an organism which causes malaria. Temperature and precipitation are the main climate factors that affect the malarial transmission potential of the mosquito population. Assessments of the potential impact of global climate change on the incidence of malaria suggest a widespread increase of risk due to expansion of the areas suitable for malaria transmission. The predicted increase is most pronounced at the borders of endemic malaria areas and at higher altitudes within malaria areas.

The changes in malaria risk must be interpreted on the basis of local environment conditions, the effects of socio-economic development and malaria control programs or capabilities. The incidence of infection is most sensitive to climate changes in areas of Southeast Asia, South America and parts of Africa. Global warming will also provide for the first time ever the right conditions for mosquitoes to breed in Southern England, Europe and the northern United States.

Plasmodium vivax

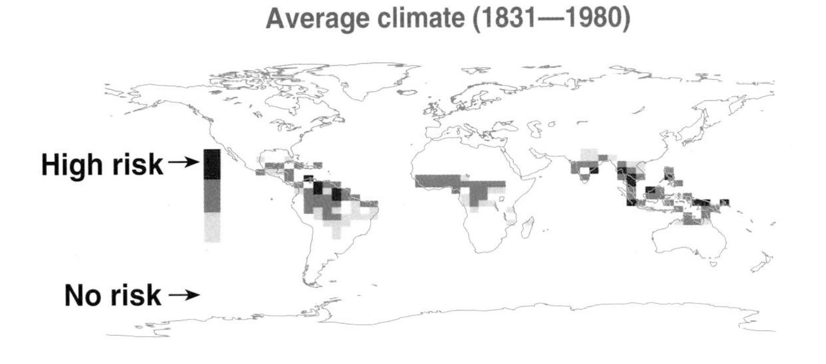

Climate change scenario (+1.2°C)

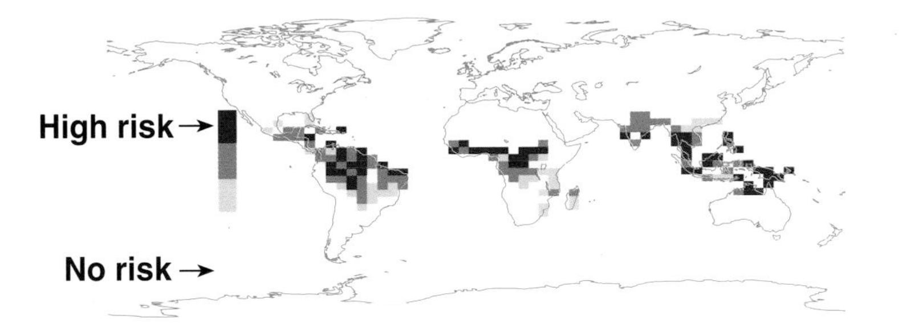

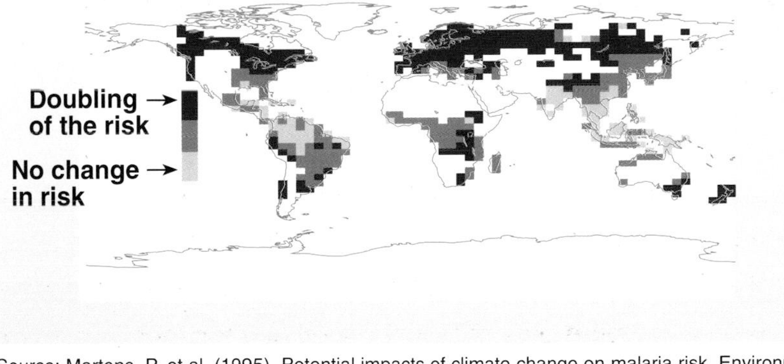

Source: Martens, P. et al. (1995). Potential impacts of climate change on malaria risk. Environmental Health Perspectives, 103(5), 458-464.

These maps illustrate where malaria has occurred in the 19th and 20th centuries and where modelers believe malaria will expand to given a moderate warming of 1.2°C in the future. The bottom map shows the change in the chance of catching malaria in the future. The chances of catching malaria in the United States will still be very low, but will be twice as great as they were during the 20th century.

Surprises

Deep Water Circulation

The circulation of the ocean is one of the major controls on our global climate. In fact the deep ocean is the only candidate for driving and sustaining internal long-term climate change, of hundreds to thousands of years, because of its volume, heat capacity and inertia. In the North Atlantic, the north-east trending Gulf Stream carries warm and salty surface water from the Gulf of Mexico up to the Nordic seas. The increased saltiness or salinity in the Gulf Stream is due to the huge amount of evaporation that occurs in the Caribbean, which removes moisture from the surface waters and concentrates the salts in the sea water. As the Gulf Stream flows northward it cools down. The combination of a high salt content and low temperature makes the surface water heavier or denser. When it reaches the oceans north of Iceland, the surface water has sufficiently cooled that it becomes dense enough to sink into the deep ocean. The 'pull' exerted by this dense sinking maintains the strength of the warm Gulf Stream, ensuring a current of warm tropical water into the North Atlantic that sends mild air masses across to the European continent. It has been calculated that the Gulf Stream delivers 27,000 times the energy of all of Britain's power stations put together. If you are in any doubt about how good the Gulf Stream is for the European climate compare the winters at the same latitude either side of the Atlantic ocean, for example London and Labrador or Madrid and New York.

The newly formed deep water sinks to a depth of between 6500 and 11,500 ft (2000 and 3500 m) in the ocean and flows southward as the North Atlantic Deep Water (NADW). In the South Atlantic Ocean it meets a second type of deep water which is formed in the Southern Ocean and is called the Antarctic Bottom Water (AABW). This is formed in a different way to NADW. Antarctica is surrounded by sea ice and deep water forms in coast polynyas or large holes in this sea ice. Out-blowing Antarctic winds pushing sea ice away from the continental edge produces these holes. The winds are so cold that they super cool the exposed surface waters. This leads to more sea ice formation and salt rejection (no one has yet been able to produce a salty ice cube!), producing the coldest and saltiest water in the world. AABW flows around the Antarctic and penetrates the North Atlantic, flowing under the warmer and thus somewhat lighter NADW and also the Indian and Pacific Oceans.

This balance between the NADW and AABW is extremely important in maintaining our present climate, as not only does it keep the Gulf Stream flowing past Europe but it maintains the right amount of heat exchange between the Northern and Southern hemispheres. Scientists have shown that the circulation of deep water can be weakened or 'switched off' if there is enough input of fresh water to make the surface water too light to sink. There is already evidence that global warming is causing significant melting of the polar ice caps, which will lead to lots of fresh water being added to the polar oceans. Global warming could therefore cause the collapse of NADW, weakening the warm Gulf Stream. This would cause much colder European winters, stormier conditions and more severe weather. However, the influence of the warm Gulf Stream is mainly in the winter so it does not affect warmer temperatures. If the Gulf Stream fails, summer temperatures will not be affected and hence global warming would still cause them to increase.

But a counter scenario is if the Antarctic ice sheet starts to melt significantly before the Greenland and Arctic ice, when things could be very different. If enough meltwater is put in the Southern Ocean then AABW will be severely curtailed. Because the deep water system is a balancing act between NADW and AABW, if AABW is reduced then the NADW will increase and expand. The problem is that NADW is warmer than AABW, and because liquid that is heated expands the NADW will take up more space. So any increase in NADW will mean a massive increase in sea level, of up to 8 ft (2.5 m). The problem is we have no idea how much fresh water it will take to shut off either the NADW or the AABW; nor can we predict which will melt first, the Arctic or Antarctic. We do know that these events have happened in the past. But if global warming continues, sometime in the future the options will be either severe alteration of European climate or an 8 ft (2.5 m) sea level rise.

Gas Hydrates

Below the world's oceans and permafrost lurks a deadly threat, gas hydrates. This is a mixture of water and methane, which is sustained as a solid at very low temperatures and very high pressures. Gas hydrates are a solid composed of a cage of water molecules, which hold individual molecules of methane. The methane comes from decaying organic matter found deep in ocean sediments and in soils beneath permafrost. These gas hydrate reservoirs are extremely unstable, as a slight increase in temperature or decrease in pressure can cause them to destabilize and hence pose a major risk.

Great Ocean Conveyor Belt

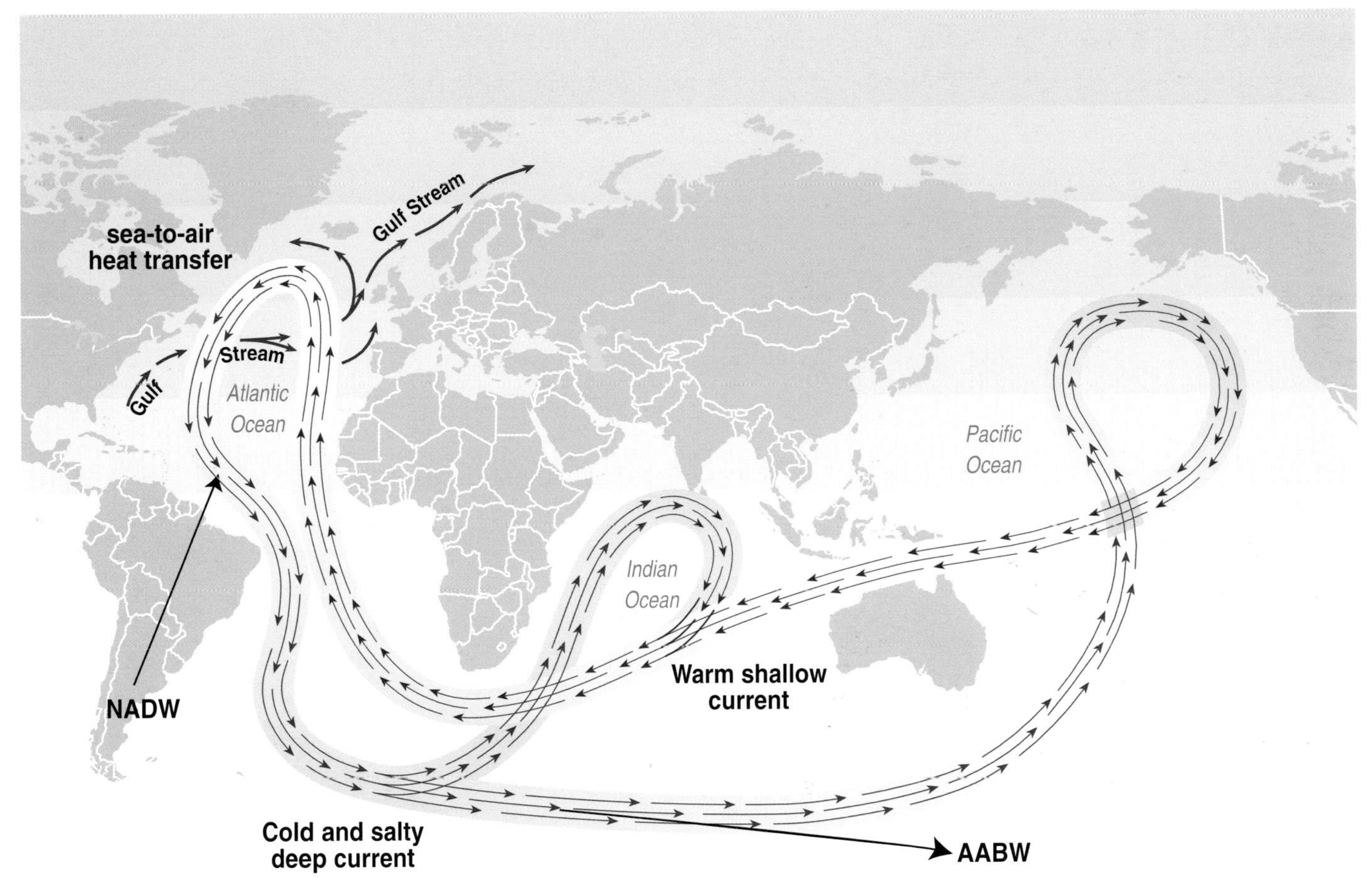

Source: Broecker, 1991, in Climate change 1995, Impacts, adaptations and mitigation of climate change: scientific-technical analyses, contribution of working group 2 to the second assessment report of the intergovernmental panel on climate change, UNEP and WMO, Cambridge University Press, 1996.

NADW: North Atlantic Deep Water

AABW: Antarctic Bottom Water

KINGS ARMS
KINGS ARMS
WATERFRONT
SEAFOOD SPECIALISTS
TAJ MAHAL
INDIAN CUISINE

Global warming will heat up both the oceans and the permafrost and could cause the gas hydrates to break down, pumping out huge amounts of methane into the atmosphere. Methane is a very strong greenhouse gas, 21 times more powerful than carbon dioxide. If enough is released it would raise temperatures even more, releasing still more gas hydrates, and producing a runaway greenhouse effect. There are 10,000 gigatons of gas hydrates stored beneath our feet compared with only 180 gigatons of carbon dioxide currently in the atmosphere. The reason why scientists are so worried about this is because there is evidence that a runaway greenhouse effect occurred 55 million years ago. During this hot-house event only 1200 giga tons of gas hydrates were released but it accelerated the natural greenhouse effect, producing an extra 9°F (5°C) of warming. One hope is that the current global warming will not be sufficent to warm the deep ocean enough to cause the release of gas hydrates.

There is a second possibility. If large parts of Greenland and Antarctica melt, a huge amount of weight is removed. The weight of ice has been pushing down the continents and when it is removed the continents can rise back to their original height. The British Isles are still recovering from the last ice age and Scotland is moving upwards while England is moving downwards. But this will mean the relative sea level around the continental shelf will become less, removing the weight and thus the pressure on the deep sea sediment. Pressure removal is a much more efficient way of destabilizing gas hydrates and hence huge amounts of methane could be released from around the Arctic and Antarctic.

When gas hydrates break down they can do so explosively. There is clear evidence that in the past, gas hydrate release has caused enormous slumping of the continental shelf and massive tsunamis, giant waves. The most famous is the Norwegian Storegga slide which occurred about 8000 years ago. It was the size of Wales and produced a 50 ft (15 m) tsunami, which is about the same size as the 2004 26th December tsunami in the Indian Ocean, which killed over 150,000 people. Hence we cannot rule out the fact that global warming could lead to an increased frequency of massive killer waves 50 ft (15 m) high hitting our coasts. Up to now only the countries around the Pacific rim which are at risk from earthquakes are prepared for this type of event, as many of these tsunamis are set off by earthquakes. But gas hydrate tsunamis could occur anywhere in the ocean.

The Amazon Rainforest

In 1542 Francisco de Orellana led the first European voyage down the Amazon River. Not only did this intrepid voyage give the Amazon River its name, from the women warriors who were encountered, but started an almost mystical awe of the greatest river and rainforest in the world. The Amazon River is a product of the Amazon monsoon, which every summer brings huge rains. This also produces the spectacular expanse of rainforest, which supports an area of the highest diversity and largest number of species in the world.

The Amazon rainforest is also important when it comes to the future of global warming, as it is a huge natural store of carbon. Until recently it was thought that an established rainforest such as the Amazon had reached maturity and thus could not take up any more carbon dioxide. Experiments in the heart of the rainforest have shown this to be wrong: it is sucking up around 5 tons of atmospheric carbon dioxide per 2½ acres (1 ha) per year. This is because plants react favorably to increased carbon dioxide, the more of it the better, as it is the raw material for photosynthesis. So having more carbon dioxide in the atmosphere acts like a fertilizer, stimulating plant growth. Because of the size of the Amazon rainforest it seems that it presently is taking up about three quarters of the world's car pollution. But things could change; global climate models suggest that by 2050 global warming could have increased the winter dry season enough that the Amazon rainforest could no longer survive. The extended dry periods would lead to forest fires, returning the carbon stored in the rainforest into the atmosphere and accelerating global warming. The rainforest would be progressively replaced by savannah, grasslands which have a much lower carbon storage potential. The Amazon rainforest at the moment is helping to reduce the amount of pollution we put into the atmosphere, but ultimately may cause global warming to accelerate at an unprecedented rate.

The Amazon River discharges 20 per cent of all the fresh water carried to the oceans.

The Amazon Basin covers 2.7 million square miles, much of which is covered with rainforest, which is under threat from deforestation. This false-color Landsat satellite image shows the devastation of the forest, with dark green being natural forest in contrast to the pale green and pink of the leveled forest.

What Can We Do?

Have the Climate Talks Failed?

The most logical approach to the global warming problem would be to significantly cut emissions, but this has major implications for the world economy. The United Nations Framework Convention on Climate Change (UNFCCC) was created at the Rio Earth Summit in 1992 to try and negotiate a world-wide agreement to reduce greenhouse gases and limit the impact of global warming.

Two major steps forward have been achieved in the last ten years. The first occurred at midnight on 13 December 1997 when the Kyoto Protocol was formed, which stated the general principles for a world-wide treaty on cutting greenhouse emissions. The Kyoto Protocol stated that all developed nations would aim to cut their emissions by 5.2 per cent of their 1990 levels between 2008 and 2012. However some countries have continued to increase their emissions significantly since 1990. The United States now produces 30 per cent more carbon dioxide pollution than in 1990, so if it were to agree with and ratify the Kyoto Protocol it would have to cut its emissions by over a third, which would be damaging to the economy.

The second breakthrough was in Bonn on 23 July 2001, when over 180 countries agreed to the Kyoto Protocol, making it a legal treaty. However, the United States, under the leadership of President George W. Bush, had already withdrawn from the climate negotiations in March 2001 and thus did not sign the Kyoto Protocol at the Bonn meeting. With the United States producing about a quarter of the world's carbon dioxide pollution this is a big blow for the treaty. Moreover the targets set by the Kyoto Protocol were reduced during the Bonn meeting to make sure Japan, Canada and Australia would join. The targets for the 37 richest and most developed countries will be about a 1 to 3 per cent cut compared with their 1990 levels. But the treaty does not include under-developed countries. This is worrying because if countries such as India and China continue to develop they will produce huge amounts of pollution. If these two countries achieve the same car-to-family ratio as Europe then there will be an extra billion cars in the world.

The third breakthrough was the Kyoto Protocol which came into force on 16 February 2005, which

An international agreement is required if we are to reduce the amount of fossil fuels we burn.

could only come into effect after Russia ratified the treaty, thereby meeting the requirements that at least 55 countries representing more than 55 percent of the global emissions passed it into national law.

So what have these nations signed up for? The 38 industrialized nations have agreed to binding targets to reduce their greenhouse gas emissions. The EU will immediately start turning the treaty into law for all member countries, forcing a cut in greenhouse gas emissions of 8 per cent on 1990 levels by 2010. The United Kingdom legal target will be 12.5 per cent to allow poorer EU countries room for development. The industrialized world will provide £350 million ($500 million) of new funds a year to help developing countries adapt to climate change and to provide new, clean technologies. Industrial countries will be able to plant forests, manage existing ones and change farming practices and thereby claim credit for removing carbon dioxide from the atmosphere. An international trade in carbon may even be started. Companies saving carbon by building clean technologies in other states will be able to claim credits which can be sold as tonnes of carbon saved on an international commodity market. This will most likely be set up in London. If countries fail to reach the first set of targets by 2012 they will have to add the shortfall to the next commitment period plus a 30 per cent penalty. They will also be excluded from carbon trading and be forced to take corrective measures at home.

The overall costs of climate change should be also be considered to investigate whether it is worth trying to do something about it. At lower levels of global warming, estimates are ambiguous about whether benefits or costs will emerge overall, with some regions showing aggregate costs while other regions will experience benefits. However, as global warming increases beyond 2-3°C above 1990 global mean temperature levels, costs emerge in all world regions. Very few studies are available to look at higher levels of warming but the literature centring on 2-3°C indicates costs on the order of half to a few percent of World GDP. Costs of climate at higher levels of warming would be higher than this, but no one knows how much. Also surprise events or abrupt change, which could be triggered by rapid rates of warming, would substantially increase the costs of climate change. Costs are expected to be unevenly distributed around world regions with damages as a share of GDP in developing countries double or triple those in developed countries, with numbers estimated as high as 5% of developing (poor) countries GDP. This is mainly due to the lack of adaptive capability of the economies of developing countries rather than to differences in the distribution of physical impacts. But as usual the poorest countries and people of the world will suffer most.

Adapt

The most sensible approach to preventing the worst effects of global warming would be to cut carbon dioxide emissions. Scientists believe a cut of between 60 to 80 per cent is required to avoid the worst effects of global warming. But the ratification of the Kyoto Protocol at the Bonn meeting in July 2001 will only amount to a cut of between 1 and 3 per cent. So the second major aim of the IPCC is to study and report on the potential sensitivity, adaptability and vulnerability of each national environment and socioeconomic system, because if we can predict what the impacts of global warming are likely to be then national governments can take action to mitigate the effects. For example, if flooding is going to become more prevalent in Britain then damage to property and loss of life can be prevented with strict new laws which limit building on flood plains and vulnerable coasts.

Wind power: a clean energy source.

The IPCC believes there are six reasons why we must adapt to climate change. 1) Climate change cannot be avoided. 2) Anticipatory and precautionary adaptation is more effective and less costly than forced last minute emergency fixes. 3) Climate change may be more rapid and more pronounced than current estimates suggest and unexpected events as we have seen are more than just possible. 4) Immediate benefits can be gained from better adaptation to climate variability and extreme atmospheric events, for example with the hurricane risk, strict building laws and better evacuation are examples of this. 5) Immediate benefits also can be gained by removing maladaptive policies and practices, for example building on flood plains and vulnerable coastlines. 6) Climate change brings opportunities as well as threats. Future benefits can result from climate change. The IPCC has provided lots of ideas of how one can adapt to climate change.

The major threat from global warming is its unpredictability. Humanity can live in almost any extreme of climate, from deserts to the Arctic, but only because we can predict what the extremes of

Alternative sources of energy are essential if we are to cut our reliance on fossil fuels. In Iceland, hot molten rock is close to the surface; water is passed close to this source of geothermal heat, superheated and used to generate both electricity and hot water for heating buildings. After use the water is still at a temperature of 40°C, and is released into large pools, that double as open air swimming pools, such as this one, The Blue Lagoon, Reykjanes.

climate will be. So adaptation is really the key to dealing with the global warming problem, but it must start now, as infrastructure changes can take up to 50 years to implement. Adaptation requires money to be invested now; and many countries just do not have the money. Also, people are often unwilling to pay more taxes to protect themselves in the future, as most people live for today.

Conclusions

So there is some good news concerning global warming. For the first time some 185 nations have made a legally binding agreement to cut emissions. It may be a very small amount that they have agreed to but it is at least in the right direction. Moreover the political pressure on the United States will be so great that many experts are sure they will finally join the treaty particularly after the battering they received from Hurricane Katrina. After ten years the scientific case is so strong for global warming that the debate about 'will it'–'won't it' happen seems to be at an end. The 2001 IPCC reports will be seen as the turning point in the global warming debate. One of the most extraordinary sources of support has been big business. With the exception of some U.S. oil companies, the business community is reacting rapidly to the threat of global warming. In the last five years companies like Ford, and oil companies like BP and Shell, have begun to pour billions into research in new technologies. Wind power is now mainstream, solar power is in rapid development, hybrid cars are on the roads. In addition there is growing support in Europe and USA for the UN Millennium Development Goals which aim to cut world poverty, increase development aid, tackle HIV/Aids, and achieve universal primary education within a sustainable development remit. This involves dealing with climate change as it firstly makes global poverty worse; as there is a direct link between development and vulnerability to natural disasters, basically the poorer you are the more like you are to be killed by a natural disaster. Second as countries develop they will burn more and more fossil fuels accelerating global warming. So different sustainable ways of development must be found to cut our dependence on fossil fuels.

Global warming is one of the greatest threats to face humanity. It is also humanity's greatest challenge. Human ingenuity is such that finding alternatives to fossil fuel is within our current technology. All that is needed is strong global political support to get something done. If we do not then the consequences will be disastrous and it will be the poorest people in the world who will suffer, the numbers of whom will be measured in billions.

Index

Recommended Reading

Houghton *et al* (editors), *Climate Change 2001: The Scientific Basis*, Contribution of Working Group I to the Third Assessment Report of the Intergovernmental Panel on Climate Change (IPCC), Cambridge University Press, 2001.

Leggett, J., *Global Warming; The Greenpeace Report*, Oxford University Press, 1990.

Harvey, L.D.D., *Global Warming: The Hard Science*, Prentice Hall, 2000.

Climate Change Impacts on the United States, Overview, National Assessment Synthesis Team, Cambridge University Press, 2000.

Metz *et al* (editors), *Climate Change 2001: Mitigation*, Contribution of Working Group III to the Third Assessment Report of the Intergovernmental Panel on Climate Change (IPCC), Cambridge University Press, 2001.

Houghton, J. *Global Warming; The Complete Briefing*, Cambridge University Press, 1994.

McCarthy *et al* (editors), *Climate Change 2001: Impacts, Adaptation, and Vulnerability*, Contribution of Working Group II to the Third Assessment Report of the Intergovernmental Panel on Climate Change (IPCC), Cambridge University Press, 2001.

Maslin Mark, *Global Warming, A Very Short Introduction*, OUP, 2004

Weart, Spencer R. The Discovery of Global Warming : (New Histories of Science, Technology, and Medicine) Harvard University Press 2003

Biographical Note

Professor Mark Maslin is at the Environmental Change Research Centre, Department of Geography, University College London. He is a leading climatologist with particular expertise in past global and regional climate change. He has published over 75 papers in journals such as *Science* and *Nature*, written five popular books, and many popular articles for publications such as *New Scientist* and *Guardian*. Mark Maslin has appeared on radio, television and is consulted regularly by the BBC on issue of climate change.